Deep-Sky Companions

The
Messier
Objects

For Donna,

the

center

of my

universe

Deep-Sky Companions

The Messier Objects

Stephen James O'Meara

Foreword by David H. Levy

SKY PUBLISHING CORPORATION

CAMBRIDGE
UNIVERSITY PRESS

PUBLISHED BY THE PRESS SYNDICATE OF THE UNIVERSITY OF CAMBRIDGE
The Pitt Building, Trumpington Street, Cambridge CB2 1RP, United Kingdom
AND BY SKY PUBLISHING CORPORATION
49 Bay State Road, Cambridge, Massachusetts 02138-1200, USA

CAMBRIDGE UNIVERSITY PRESS
The Edinburgh Building, Cambridge CB2 2RU, United Kingdom
40 West 20th Street, New York, NY 10011-4211, USA
10 Stamford Road, Oakleigh, Melbourne 3166, Australia

First published 1998

Printed in the United States of America

Typeset in Utopia Regular 9/13pt   [SE]

A catalogue record for this book is available from the British Library

2658

Library of Congress Cataloguing in Publication data

O'Meara, Stephen James, 1956–
     The Messier objects field guide : a new look at the most famous
deep-sky wonders in the heavens / Stephen James O'Meara : foreword
by David H. Levy.
       p.   cm.
     Includes index.
     ISBN 0 521 55332 6
     1. Astronomy–Charts, diagrams, etc.   2. Astronomy–Observers'
manuals.   3. Galaxies–Charts, diagrams, etc.   4. Stars–Clusters–
Charts, diagrams, etc.   5. Nebulae–Charts, diagrams, etc.
6. Messier, Charles. Catalogue des nébuleuses et amas d'étoiles.
I. Title.
QB65.O44   1997
523.8–dc21 96-51773 CIP

ISBN 0 521 55332 6 hardback (Cambridge University Press)
ISBN 0 933346 85 9 (Sky Publishing Corporation)

Contents

Foreword by David H. Levy vii

Preface ix

Acknowledgments xiii

1 **Charles Messier and his catalogue** DAVID H. LEVY **1**

2 **How to observe the Messier objects** 9

3 **The making of this book** 25

4 **The Messier objects** 39

5 **Some thoughts on Charles Messier** 285

6 **Twenty spectacular non-Messier objects** 289

APPENDIX A
Objects Messier could not find 297

APPENDIX B
Messier marathons 299

APPENDIX C
A quick guide to navigating the Coma–Virgo cluster 301

APPENDIX D
Suggested reading 303

Foreword

A fragile observing log, yellowed with age, lay protected under glass in the main hall of the Paris Observatory, a room crowded with old telescopes that whispered memories of starlit nights long past. On the log's pages, in words written more than two centuries earlier, the great comet hunter Charles Messier commented on the weather, sky conditions, how his observing sessions went, and his plans for a new telescope. The thought occurred to me that Times really haven't changed. The concerns he had then are my concerns today, and the feeling he must have had as he peered through his telescope at some unknown fuzzy object, his brief moment of tension – could it be a comet? – is a feeling I know well from my own comet searches. As I read Messier's words, the years began to fall away. There he was, surveying the sky from the tower of the Hôtel de Cluny, sorting out comets from unknown nebulae and star clusters, and there I was right next to him.

Charles Messier's story is a prime example of serendipity in astronomy: we remember him today more for his catalogue of noncometary celestial objects than for the comets he discovered. Yet Messier was the first person to discover comets as part of an organized search for them. To optimize his search efforts he kept a list of nebulous objects that merely masquerade as comets so that he and others would not be fooled by them more than once. The 12 comets he is credited with discovering are all long gone, but his catalogue of more than 100 deep-sky treasures remains a collection of some of the finest showpieces in the heavens.

My own hunt for the Messier objects began in August 1962, with Echo, my 3-inch reflector telescope, and the Andromeda Galaxy as my first quarry. Using a nearby 4th-magnitude star coupled to two fainter stars as an arrow, I moved the telescope gently across two or three fields of view. Suddenly my eye caught sight of a bright patch of fuzzy light. There it was, my first Messier object, right where it belonged. It may have looked like only a cloud in the sky, but I would read that its light had left that galaxy more than a million years ago (today we know it's more than 2 million). And that made the experience one I would never forget.

There can be no better exercise for a beginning astronomer than to find and observe all of the Messier objects. It is the best way to get acquainted with the riches the night sky has to offer. The stunning beauty of the Orion Nebula (Messier 42), the delicate wispy structure of the Omega Nebula (Messier 17), and the vastness of the Andromeda Galaxy

(Messier 31) just hint at the grandeur represented by these gems of deep space.

Charles Messier would have been proud to know the author of this book. Without a doubt, Steve O'Meara is one of the best visual observers in the world today. In 1985 he became the first person to see Halley's Comet since 1911 – a full seven months before anyone else saw it. (Messier himself had wanted to be the first to herald Halley's Comet's 1758 return, but another observer beat him to it.) Steve's acute planetary observations are highly regarded, and his intricate drawings of deep-sky objects are superb. Steve has written a book that will guide you through the sky to meet the Messier objects one by one. Finding them is both fun and challenging; this book will turn that challenge into a memorable, and perhaps even a lifelong voyage of discovery.

David H. Levy

Preface

Two and a half centuries ago a French comet hunter named Charles Messier began compiling a catalogue of nebulous sky objects. He explained his motivation in the French almanac *Connaissance des Temps* for 1801:

> What caused me to undertake the catalogue was the nebula I discovered above the southern horn of Taurus on 12 Sept. 1758, while observing the comet of that year. . . . This nebula had such a resemblance to a comet, in its form and brightness, that I endeavored to find others, so that astronomers would not confuse these same nebulae with comets just beginning to shine.

Messier died long before twentieth-century astronomers realized the profound nature of these hauntingly diffuse glows. The 110 "Messier objects," it turns out, are an eclectic collection of celestial treasures: 39 galaxies, 57 star clusters, 9 nebulae, a supernova remnant, a swath of Milky Way, a tiny grouping of stars, a double star, and even a duplication. The list includes the most massive and luminous galaxy known, the ghostly remains of a cataclysmic stellar explosion, and an immense cosmic cloudscape that cradles newborn stars in dense cocoons of hydrogen gas.

Every Messier object is within reach of a small telescope, and many are visible with binoculars and the naked eye, especially under clear, dark skies far away from city lights. Amateur astronomers of all ages enjoy tackling the Messier catalogue members, because they represent a good sampling of what's "out there," and because finding them helps to hone observing skills. In a sense, the Messier objects are the testing grounds for budding skywatchers.

The Messier objects entered my life in 1966, when I was 10 years old. They were mentioned in the *Sky Observer's Guide*, a Golden Guide by R. Newton Mayall and Margaret Mayall, whose words also taught me the basics of astronomy. I quickly located the brightest Messier objects from my back porch in Cambridge, Massachusetts. Then one winter evening a friend loaned me his 2-inch refractor. I recall pointing its white enameled tube over a frozen city landscape, beyond the smoking chimneys, and seeing swirls of nebulosity surrounding the Trapezium, a group of young, bright stars in the mighty Orion Nebula (Messier 42, or M42), then the knitted stars of the Pleiades (M45) – Lord Tennyson's "glittering swarm of fireflies tangled in a silver braid." These views were visual poetry, and like Tennyson and countless others before him, I became captivated by the allure of the stars.

Several years later, when I acquired a 4½-inch reflector, another friend gave me a box of *Sky & Telescope* magazines – a monthly publication devoted to the hobby of amateur astronomy. An article immediately caught my eye; it was one in a series by John Mallas and Evered Kreimer spotlighting the Messier objects. Each article featured drawings by Mallas and photographs by Kreimer, as well as visual descriptions and brief histories of the objects. On the first clear night I eagerly set up the scope, and with magazines and red flashlight in hand, I followed along with Mallas and Kreimer as they toured the "M" objects. No longer feeling alone in my pursuits, I began dedicating one clear night to each Messier object – studying it, writing down impressions, and making drawings.

That Mallas and Kreimer series was later compiled into a book called *The Messier Album*, which served the astronomical community for many years to follow. But time has marched on. The optical quality of many commercially available telescopes and eyepieces is superior to the telescope Mallas used a quarter-century ago, providing sharper images and revealing fainter details. We also have more accurate astronomical information today about many of the objects – their sizes, distances, magnitudes, and more – than we did when *The Messier Album* was written. Object positions and maps have been updated to equinox 2000.0 coordinates from the equinox 1975.0 coordinates used by Mallas and Kreimer. Clearly, the time was right for a fresh new look at these classical astronomical specimens.

As observers today, we are not only better equipped but also wiser. Looking back, it is hard to believe that Mallas used a 4-inch f/15 refractor to make his observations. Although a popular telescope type in its day, it was best suited for studying the moon and planets. Today's deep-sky observers prefer rich-field telescopes over long-focus refractors. It is also amazing that Mallas made his observations from a Los Angeles suburb tainted by light pollution and smog! These problems are even more pervasive today and represent an insidious threat to our continued enjoyment of the heavens. More and more, residents of cities and suburbs must pack their gear and drive some distance from their homes to less populated areas to enjoy a night under the stars. And several times a year thousands of amateur astronomers journey to national conventions and star parties held at first-class, dark-sky sites. In fact, one such journey inspired this book.

For a week in May 1994, I attended the Texas Star Party, a deep-sky observing event held on the Prude Ranch near Fort Davis, Texas, at an altitude of just over 5,000 feet. One sparkling-clear night, Al Nagler, founder of Tele Vue Optics, showed me a wide-field view of the Milky Way – specifically, the region known as M24 – through his 4-inch Genesis refractor. The field was bristling with starlight and threads of dark nebulosity. I didn't want to take my eye from the telescope. Later that week, I borrowed another friend's Genesis and spent three hours studying the Whirlpool

Galaxy (M51) in Canes Venatici; I spent many more hours on subsequent nights. The graceful spiral arms, the numerous punctuating knots – all the subtle detail was awesome to behold. I began to wonder what the other Messier objects would look like through a telescope of uncompromising quality, with superior eyepieces, and viewed from the darkest sites on earth. Thus, I decided to revisit, one by one, the deep-sky gems that Charles Messier catalogued more than two centuries ago, and that started me on what was to become my latest adventure in an already long and exciting career in astronomical observation, teaching, and writing. The result is this book, which I hope will inspire and inform you as much as Mallas and Kreimer's seminal articles and book did for me and the countless others who have used them.

The purpose of this book is to provide new *and* experienced observers with a fresh perspective on the Messier objects. Chapter 1 is a brief account by world-renowned comet discoverer David Levy of the life of Charles Messier, a comet hunter himself, and how his catalogue of deep-sky curiosities came to be. Chapter 2 introduces beginners to the basics of skywatching and to some important terms and concepts. It is designed specifically to help newcomers orient themselves to the sky and start locating the brightest Messier objects. In chapter 3, "The Making of This Book," I review the methods and the equipment with which I conducted the observations described herein and provide additional information about the book's content.

Of course, the heart of this book is chapter 4, which looks in detail at each Messier object. Since this book is a "companion," I've used a conversational tone; I speak to you as if I'm with you in the field. Along with the descriptive text, I have provided for each object a list of essential data including its coordinates, size, brightness, and distance. The equinox 2000.0 finder charts have been carefully drafted to work together with Wil Tirion's all-new, wide-field constellation map at the back of the book; together they will enable you to quickly zero in on your targets. A new and comprehensive translation of Messier's original published catalogue was commissioned and is included here. It supersedes earlier translations, which often were abridged and prone to occasional errors and misinterpretations. I have also included, in appendix A, the endnotes to Messier's original catalogue, in which he lists a number of objects reported by other observers that he tried, but failed, to find himself.

The detailed drawings I made of the Messier objects are another distinctive feature of this book. Each drawing was based on several hours of observing each object over several extremely transparent nights. I think you will find these illustrations revealing and useful in helping you to see subtle details in the objects that you may not have noticed before (and which may not be apparent in photographs). I have also updated and revised many of the objects' magnitude estimates, offered thoughts about some of the "missing" Messier objects, and distributed observing challenges throughout the chapter.

In chapter 5, Some thoughts on Charles Messier, I offer some summarizing thoughts or "analysis" on Messier and his catalogue, which I felt compelled to do having spent so much time thinking about the man and what he saw – and didn't see. I also felt compelled to describe, even in a book devoted to the Messier catalogue, 20 of my favorite *non*-Messier objects, in chapter 6. Consider them honorable mentions to the catalogue that are conspicuous by their absence from it and certainly deserving of a look while you're out hunting Messier objects.

The appendices contain additional information that you will find useful, including a brief discussion about "Messier marathons," a guide to navigating the Coma–Virgo Cluster, and a suggested reading list.

Perhaps the most unique aspect of this book, and what I most want to convey, is the *approach* I take to observing. It's an approach based on creative perception and on using the imagination to see patterns and shapes in the subjects seen through the eyepiece. It involves using not just your eye but also your *mind's eye* to associate those patterns and shapes with things that are familiar to you, to create pictures and even stories. Rather than barrage you with just facts (of which you get plenty), I thought you'd also enjoy seeing these objects in new ways – especially the clusters, whose multitude of inherent shapes lend themselves to being seen as celestial Rorschach tests. By using the imagination you can add another dimension to your observing – a highly personal and entertaining one (after all, this is a hobby). Anyone who has read *Hard Times* by Charles Dickens will understand my protest of a diet consisting totally of fact. (By teaching youngsters fact not fancy, conformity not curiosity, Dickens's bleak character Thomas Gradgrind tried to stifle inquisitive minds.) If you have never gazed at the ethereal quality of a Messier object through a telescope, I encourage you to look upon them as you would a painting or a piece of art – and let that art add meaning to your experience.

It is my hope that this book will not only introduce you to the objects themselves – or reintroduce you, as the case may be – but that it will also challenge you to raise your observing skills to a higher level and to push your visual limits. I hope it compels you to search for new and mysterious aspects about these objects, to see them in rich and creative ways, and to grow as an observer.

I know the magic of the Messier objects because I have been under their spell for three decades. Today I see the same magic in my wife's eyes, whenever she raises her binoculars to the sky and happens upon her "comets." May the spell never be broken.

Stephen James O'Meara
Volcano, Hawaii

Preface

Acknowledgments

"If I have seen so far, it is because I have stood on the shoulders of giants." Like Sir Isaac Newton and others before and after him who used this axiom, I would like to recognize the giants who have helped me in my observational and literary journeys.

Highest tribute goes to the late Walter Scott Houston, who shared his observing experiences and techniques in his Deep-Sky Wonders column in *Sky & Telescope* magazine for so many years. I will never forget the times we spent by the campfire at the annual Stellafane Convention on Breezy Hill in Vermont, or in chaise longues at the Winter Star Party in the Florida Keys, just gazing at the stars and musing on the limits of vision.

I am honored to recognize George Phillips Bond (1825–1865), second director of Harvard College Observatory, whose dedication to unlocking the visual mysteries of the Orion Nebula with the Harvard Observatory's 15-inch refractor led ultimately to his premature death. Reading his diaries two decades ago kept me enchanted on many a cloudy night, and taught me how to be a patient and persistent observer.

A deep bow to "envelope pushers" Barbara Wilson and Larry Mitchell, who roped and tied this wild planetary observer at the Texas Star Party and force-fed me deep-deep-deep-sky objects until I became a convert – thank you (I think)! Assisting them was a phalanx of galaxy hunters, including David Eicher, Tippy and Patty D'Auria, and Jack Newton. Brian Skiff introduced me to many visual challenges, including seeing faint globular clusters with the naked eye. Peter Collins first introduced me to the more challenging Messier objects. David Levy was always around to say, "No, Steve, that's a galaxy not a comet." And at several star parties, Tom and Jeannie Clark were incredibly generous with their Tectron telescopes, encouraging me to sweep the Coma Cluster with these enormous Dobsonians until I nearly fainted with delight.

Al Nagler and the editors of *Sky & Telescope* will never fully know how grateful I am to them for helping me to complete this journey. Thanks for your special encouragement and support.

Steve Peters gets the blue ribbon for nurturing the book idea, and kudos to Simon Mitton and Cambridge University Press for publishing the work. Special thanks to Lee Coombs, Martin Germano, Chuck Vaughn and George Viscome for their stunning astrophotographs. I greatly appreciated the expert assistance of Brent A. Archinal, Kevin Krisciunas, Larry Mitchell, Brian Skiff, and Barbara Wilson, who reviewed drafts of the text and made necessary corrections and welcome suggestions. Thanks to

Steve and Tom Bisque of Software Bisque for their generous donation of *The Sky* astronomy software, which was used in creating the finder charts, and to master astronomical cartographer Wil Tirion for the wide-field map of the Messier objects at the back of the book. Thanks also to my good friend Storm Dunlop for his excellent translation of Messier's catalogue from the original French.

Kudos to Nina Barron for her expert, sensible, and sensitive copy-editing of the manuscript. Heartfelt appreciation to the Dillingham family – Ken, Lina, Serena, and Karen – for the use of their ranch in the saddle of Kilauea and Mauna Loa volcanoes, where clear skies, steady seeing, and the feeling of home helped me to finish the observations on time. Thanks also to Nina Barron for proofreading the manuscript.

And no words can express the love and devotion I have for my wife, Donna, who, despite having been "husbandless" for a year, found solace in exploring erupting volcanoes, practicing her "free dives" to unknown depths, and peering inquisitively into the eyes of the night.

Finally, I would like to thank our "children," Pele-Hiiaka of Volcano, Milky Way, and Miranda-Pywacket for keeping their digits off the keyboard when I wasn't looking. Alas, I cannot blame them if any errors have sneaked unannounced into this book; I take full responsibility.

DAVID H. LEVY

1 Charles Messier and his catalogue

In 1744 a brilliant comet punctuated the night sky, attracting the eyes of people around the world and capturing the imagination of a 14-year-old named Charles Messier. Seven years later the young man moved from his native Lorraine, a region of southern France, to Paris to find his fortune, which he felt was out among the stars. The key to his success, he would soon discover, was in hunting for comets.

If Messier's competitive spirit was anything like that of today's comet hunters, there is a good chance that it was born out of failure, not success. In 1758 many people were rushing to find the comet whose return Edmond Halley had predicted years earlier, and the keen-eyed Messier had a good chance to be the first. At the time, he was clerk at the Marine Observatory, which the astronomer Joseph Nicolas Delisle had established a decade earlier at the Hôtel de Cluny in Paris. Delisle, who was nearing 70, had taken Messier under his wing in 1751 and had begun to shape his career. By 1754 Messier had become a highly skilled and respected observer. And his main pursuit was hunting for comets.

The aged Delisle had great hopes that his young protégé would be the first to sight Halley's Comet. Thus, Delisle had prepared a star chart with two plots of the comet (whose position he had calculated), and instructed Messier to conduct a search of these areas. Messier began his search in mid-1757; it ended successfully on 21 January 1759, though the effort took much longer than he had hoped. He explained in the *Connaissance des Temps* for 1810:

> The whole day was very fine and without cloud; in the evening I went over the sky with the telescope, keeping the limits of the two ovals drawn [by Delisle] on the celestial chart which was my guide. At about six o'clock I discovered a faint glow resembling that of the comet I had observed in the previous year: it was the Comet itself, appearing 52 days before perihelion!
>
> There is cause to presume that if M. Delisle had not made the limits of the two ovals so restricted, I would have discovered the comet much earlier.

But Messier had been beaten. Johann Georg Palitzsch, a German farmer and amateur astronomer who lived near Dresden, sighted Halley's prodigal comet on Christmas night 1758 – a month before Messier did. The news of Palitzsch's find, however, did not reach Paris for several months, so Messier was unaware that he had been "scooped." In fact, Messier

Portrait of Charles Messier (June 26, 1730 – April 12, 1817) in 1801, twenty years after his last catalogue observation on April 13, 1781. *Bulletin* de la Société Astronomique de France, 1929.

deemed his find one of the most important astronomical discoveries ever, for it showed that comets could orbit the sun and return again. Messier immediately informed Delisle, who confirmed the observation but told Messier not to announce it under any circumstances, for reasons that are unclear. Messier did not complain, at least publicly, about this strange embargo on his "discovery." Only when Messier sighted the comet again after it had rounded the sun did Delisle allow him to publish his observations.

"Such a discreditable and selfish concealment of an interesting discovery," wrote J. Russell Hind a century later, "is not likely to sully again the annals of astronomy. Some members of the French Academy looked upon Messier's observations, when published, as forgeries, but his name stood too high for such imputations to last long, and the positions were soon received as authentic, and have been of great service in correcting the orbit of the comet at this (1835) return."

Furthermore, Halley's 1758 return was not announced until 1 April 1759, three months after the first sighting and three weeks after the comet had rounded the sun. Ironically, the only vital positional observations of the comet that winter were Messier's. By this time, Messier had heard of Palitzsch's earlier sighting. Messier's disappointment possibly had something to do with his decision to start a systematic search for comets.

COMET MASQUERADERS

In conducting his searches, Messier did not have the benefit of good star charts like we have today, showing the positions of galaxies, star clusters, and nebulae – what the great twentieth-century comet hunter Leslie Peltier termed "comet masqueraders." In Messier's day, these objects were largely unknown and uncharted. Thus, he must have been surprised when, on 28 August 1758, while tracking a comet, he came across a fuzzy patch near the star Zeta (ζ) Tauri. At first he thought he'd snared a new comet. But the fuzzy patch never moved with respect to the stars, as comets always do. Realizing he had been fooled by the sky's version of a practical joke, Messier began to build a catalogue of what he called these "embarrassing objects."

That first entry in his catalogue, Messier 1, is now commonly called the Crab Nebula. Although Messier never realized it, his first object was probably the most interesting thing ever to cast light into his telescope. It is all that remains of a supernova that was observed in 1054, which shined as brightly as Venus.

By 1765, Messier had compiled a list of 41 such objects. Of those, only 17 or 18 were his own discoveries; the rest had been seen previously by others (whom he acknowledged). Before submitting the list for publication, he decided to round it out with a few more objects. So on 4 March

Facsimile of a page written by Messier in 1754, detailing solar and lunar observations. *Bulletin* de la Société Astronomique de France, 1929.

Deep-Sky Companions: The Messier Objects

of that year he determined the positions of the Great Nebula in Orion (M42 and M43), the Beehive Cluster (M44) in Cancer, and the Pleiades (M45) in Taurus. He presented his list of 45 nebulae and star clusters to the Academy of Sciences in Paris in February 1771, and it appeared in the Academy's *Memoirs* for that year, which were actually published in 1774. Now with eight comet discoveries and a fresh appointment as Astronomer to the Navy, Messier's career was rocketing. Indeed, within days after submitting his list of objects for publication, he had already discovered four more star clusters, and by April 1780 the list had grown to 68 objects. Of the 23 new objects – most of which he found while observing comets – some had been recorded previously by Messier's contemporaries, including his younger colleague Pierre Méchain (who discovered 32 new nebulous objects between 1780 and 1781), the Italian observer Barnabus Oriani, and the French astronomer Nicolas Louis de Lacaille. This updated catalogue was published in the French almanac *Connaissance des Temps* for the year 1783.

On 13 April 1781, less than a month after William Herschel's discovery of the distant planet Uranus, Messier observed the 100th cometlike object to be logged in his catalogue. Although he had intended that object to be the last one, he decided at the last minute to include three additional objects observed by Méchain, but which Messier did not have time (before the publication deadline) to observe himself. Thus, Messier's final catalogue described 103 objects and was published in the *Connaissance des Temps* for 1784.

In 1921 a French popularizer of astronomy, Camille Flammarion, found notes about an additional object in Messier's personal copy of the catalogue. The object (NGC 4594, the Sombrero Galaxy), in Virgo, was designated M104. In 1947 Canadian astronomer Helen Sawyer Hogg proposed adding four more objects discovered by Méchain to the catalogue. Méchain had described them in a letter to a German astronomy journal, and notations in Messier's copy of the printed catalogue suggest that he was aware of them. One of the objects was M104, so Hogg labeled the remaining three M105, M106, and M107. Subsequently, Owen Gingerich, astronomical historian at Harvard University, recommended the inclusion of two more galaxies, NGC 3556 and NGC 3992 in Ursa Major, because they too were mentioned in Messier's original catalogue; they became M108 and M109, respectively. M110 is the most recent, and presumably the last, addition to the collection. Its inclusion was suggested in 1966 by the late Kenneth Glyn Jones, who noted an engraving and descriptions by Messier of the Andromeda Galaxy (M31) and its two companions that had been published by Messier in 1807. One of the companions had been included in the original catalogue, as M32, but the other (NGC 205) had not, so it became M110. Indeed, Messier had first seen it on 10 August 1773.

Messier.

MYSTERIOUS AND "MISSING" MESSIER OBJECTS

Six objects in the Messier catalogue are particularly curious. M40 is simply two faint stars. Messier detected it while searching for a nebula described by the seventeenth-century observer Johann Hevelius. "We presume," wrote Messier, "that Hevelius mistook these two stars for a nebula." Regardless, Messier included them in his catalogue. Similarly, M73 is a small group of a few faint stars, which he thought was nebulous at first glance. The cases of four "missing" Messier objects – 47, 48, 91, and 102 – are different. No nebula or cluster resides at or near the positions recorded for them by Messier. The mysteries have been explained in different ways. In 1934 Oswald Thomas, in his book *Astronomie,* suggested that Messier 47 was actually NGC 2422, a bright open cluster in Puppis. Dr. T. F. Morris, a long-time member of the Royal Astronomical Society of Canada's Montreal Centre, independently arrived at the same conclusion in 1959, suggesting that Messier may have made a simple mistake in computing the object's position. Morris discovered that by reversing Messier's directions both in right ascension and in declination from the star 2 Puppis (the star Messier used for reference), one arrives at NGC 2422.

Gingerich believes the object that Messier listed as number 48 is probably the cluster NGC 2548, which lies about 5° south of the position that Messier documented, a conclusion reached independently by Morris and others. NGC 2548 has the same right ascension that Messier recorded, making it likely that he unwittingly used the wrong reference star when determining the object's declination.

The true identity of Messier 91 is the most perplexing, because, as Jones comments, the problem lies not in the lack of candidates but in the wealth of them. M91 is in the Coma–Virgo Cloud of galaxies, the densest region of galaxies visible in small telescopes. Some observers believe that NGC 4571, a 12th-magnitude galaxy slightly northwest of the position Messier recorded, is M91. Others argue that this particular galaxy would have been too faint for Messier to see. Gingerich, on the other hand, believes that M91 is nothing more than a duplicate observation of M58, which has a similar right ascension but is off in declination by 2¾°. Most observers, however, have come to accept a new theory. In 1969, W. C. Williams, a Texas amateur astronomer, proposed that Messier had originally offset the coordinates of M91 using M89 as a reference point but that when he plotted the object, he mistakenly chose M58 as the reference point. If that is what happened, then NGC 4548, a barred spiral galaxy, is Messier 91.

Finally, Messier 102, which Méchain first observed and whose position Messier did not check, is generally believed to be a case of mistaken identity. According to Hogg, Méchain discovered that the object designated as M102 was a duplicate observation of M101, and explained the error in a letter published in the 1786 *Berliner Astronomisches Jahrbuch,* which contained a copy of Messier's supplement. An English translation of the relevant paragraph of that letter reads:

> On page 267 of the *Connaissance des Temps* for 1784 M. Messier lists under No. 102 a nebula which I have discovered between omicron *Boötis* and iota *Draconis*: this is nothing but an error. This nebula is the same as the preceding No. 101. In the list of my nebulous stars communicated to him M. Messier was confused due to an error in the sky-chart.

"The real puzzle," notes Gingerich in *The Messier Album*, "is why this letter was overlooked by so many astronomers for so long." Not until Hogg published it in the *Journal* of the Royal Astronomical Society of Canada in 1947 was the confusion regarding M102 finally cleared up.

Messier's long career began to wane when he suffered a stroke in 1815, which left him partially paralyzed. Two years later he died at age 87. He left a wonderful legacy, not of comets, but of magnificent deep-sky objects that he sought to avoid – 109 lavish monuments to the skill of one of the most perceptive and enthusiastic observers of the night sky.

2 How to observe the Messier objects

The purpose of this chapter is to help new observers get started. (Seasoned observers may want to skip to the next chapter, though you should read the section on "Observing Tips" before you do.) It contains enough essential information to help you find a few Messier objects. There is no avoiding the need to learn sky directions, recognize the major constellations and their brightest stars, know how to use a star map, and understand how the sky changes with the seasons. You can teach yourself just about all these things with a star wheel, or *planisphere*. These special sky maps are inexpensive, fun and easy to use, and very educational. (Star wheels can be purchased at any nature store or planetarium gift shop.) To purchase the proper star wheel, you will need to know your latitude on earth so you can determine which stars are visible from your location. Consult a world atlas at your local library. For further help in finding the constellations, refer to *The Star Guide: A Unique System for Identifying the Brightest Stars in the Night Sky,* by Steven L. Beyer, which is listed in the "Suggested Reading" section.

Although anyone can enjoy this book, the following section on sky orientation and many references to sky directions throughout the book are targeted primarily for observers at mid-northern latitudes. Observers in the far south will not be able to see several of the most northerly Messier objects. Chapter 6, however, describes many southerly objects either missed or ignored by Messier. A good supplementary guide book for southern observers is E. J. Hartung's *Astronomical Objects for Southern Telescopes.* Otherwise, the material in this book, especially the descriptive notes of the Messier objects, can be used by all.

NAVIGATING THE SKY

Assuming you live in the Northern Hemisphere and have a star wheel and a red flashlight (white light ruins your night vision), the next step is to go outside and orient yourself. One easy way is to use a compass to find the cardinal directions – north, south, east, and west. Do this in the daytime, standing where you expect to set up your telescope. Select an object on the distant horizon – a tree, building, smokestack, or mountain – to mark the main cardinal points. For example, if your star wheel displays a bright star or constellation rising in the east, you know to look in the direction of, say, a willow tree, your eastern landmark.

When night falls, your first mission is to find the North Star, or *Polaris*. It will be your unfaltering guide. The North Star is very close to where earth's imaginary axis of rotation intersects the dome of the sky. Unlike other bright stars, it remains essentially in the same position every night all night as the earth turns. Despite its public reputation, the North Star is not the brightest star in the night sky; Sirius, a southern star, holds that honor. In fact, there are 48 stars brighter than Polaris.

The height of the North Star above your horizon (its *altitude*) is the same as your latitude on earth. If you lived on the North Pole (latitude 90°), the North Star would be directly overhead (altitude 90°). If you lived on the equator (latitude 0°), the North Star would be on the north horizon (altitude 0°). I live on Hawaii's Big Island (latitude 20°), so I look for a solitary yellowish star of moderate brightness 20° above my north horizon.

How high is 20°? Draw an imaginary line from the north horizon to the point directly overhead (the sky's *zenith*). That line spans an angular distance of 90°. Twenty degrees is almost one-quarter of the way from the horizon to the zenith. To measure angular distance, hold an upright fist at arm's length and look at it with one eye closed. The amount of sky covered by the fist is about 10°. For me to find the North Star from Hawaii, I would face north, place the base of my upright fist on the horizon line, make a fist with my other hand, and place it on top of the first. Two fists equal about 20°. The North Star should be sitting on the top fist. If you live in New York City (latitude 40°), the North Star will be four fists above your north horizon.

Now you can find the Big Dipper. Using your red flashlight turn your star wheel until the time of night lines up with the current date. You will be looking northward, so hold the star wheel with the north horizon down; the word "north" or "northern horizon" should be upright. Looking at your star wheel, read the position of the Big Dipper as you would the hand of a clock. For example, on 1 July at 8 p.m. the Big Dipper is to the upper left of the North Star, at the 10 o'clock position. The closest star to Polaris in the bowl of the Big Dipper is Dubhe. It lies 30° (three fists) away. If you open your hands and stretch out your fingers, the angular distance between the tip of your thumb and the tip of your little finger is approximately 20°. Stargazers have long used Dubhe and Merek (the other star marking the bowl's outer side) to point to Polaris, hence the term "pointer stars".

Compare the size of the Big Dipper in the sky with its size on the star wheel. This is an important exercise, because to find less obvious constellations, you will need to scale what you see in the sky with what you see on your planisphere, or vice versa. (Constellations appear larger when they are at or near the horizon than when overhead. It is the same optical illusion that makes the moon look bigger when it is on the horizon than when overhead.) Scaling star patterns is also necessary when you want to use the more detailed star charts in this book to find Messier objects with your telescope.

To find other star patterns, first locate the brightest stars, get a feeling for the scale, then look for the fainter stars that make up the constellations. What is wonderful about hunting Messier objects is that you do not have to know the entire sky to find or enjoy them, just like you don't have to know *everything* about New York City if you want to visit only the Empire State Building, the Statue of Liberty, and Central Park. It is the same way with this book. You can be selective and enjoy the sights you want to see. In fact, you can learn the constellations as you seek out the different Messier objects. At least one Messier object each season is bright enough to be seen with the naked eye. Furthermore, these objects are plotted on good star wheels.

TYPES OF MESSIER OBJECTS

There are three principal categories of objects in the Messier catalogue: *star clusters*, *nebulae*, and *galaxies*. Each is divided into subclasses: *open* and *globular* clusters; *diffuse* and *planetary* nebulae; and *spiral, elliptical,* and *irregular* galaxies. Many of the Messier objects inhabit our own galaxy, the Milky Way. What does our galaxy look like? That is hard to know because we live within it. It's like trying to determine the shape of a forest from a spot deep inside it. But astronomers believe it looks something like this:

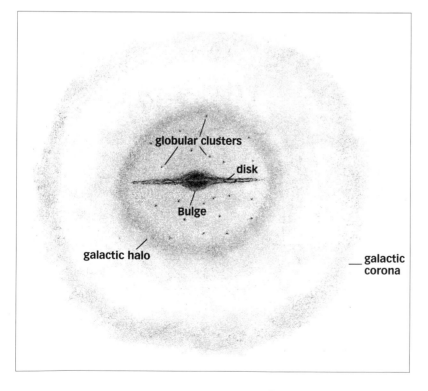

Our galaxy's *nucleus* measures about 10 light years across (1 light year is the distance that light travels in one year – 6 trillion miles!). The nucleus is surrounded by an egg-shaped bulge spanning about 32,000 light years and containing millions of stars. Encircling the *bulge* is a flat, pancakelike *disk* of dust, gas, and stars, which measures about 100,000 light years in diameter. The disk material is probably arranged in spiral arms, but no one knows for sure the shape of these arms. This entire system is contained in a spherical *galactic halo* of older stars and interstellar matter; the diameter of the halo is about 130,000 light years. Finally there is the *galactic corona* extending some 200,000 to 300,000 light years. Researchers believe that the mysterious dark matter (material that escapes visual detection; its presence is inferred from gravitational effects) inhabits this region.

The Milky Way Galaxy is our home in the universe. The universe is replete with billions of galaxies, each of which is home to billions of suns. The Messier catalogue contains 39 galaxies, all of which are millions of light years away. At a distance of 2.3 million light years, M31, the Great Andromeda Galaxy, is the closest Messier galaxy to our own; the farthest is M87, a giant elliptical system in Virgo that is a dizzying 63 million light years distant.

Note: Don't expect to see the Messier objects as bold, sharply defined masses radiating complex pastel hues. Those images are created through the magic of color photography and do not represent the way they look through a telescope, which is fuzzy and dim. But if you stop to consider how far their light has traveled before it reaches your eye, you will better appreciate their subtle, ghostly glows.

CATEGORIES OF MESSIER OBJECTS

The Messier catalogue contains five major classes of deep-sky objects: open clusters, globular clusters, diffuse nebulae, and galaxies. Open, or *galactic,* clusters are irregular conglomerations of young stars (a dozen to thousands of them) that travel in a thin disk of stars, dust, and gas apparently encircling the galactic bulge. The stars in open clusters are loosely associated and will probably disperse after making a few trips around our galaxy. Of the approximately 1,200 known open clusters in our galaxy, the Messier catalogue contains 30, including those associated with nebulosity.

Globular clusters, on the other hand, are spherical swarms of tens of thousands to perhaps a million ancient stars situated far above and below the galactic disk. These tightly packed huddles of stars move in giant elliptical orbits in our galaxy's spherical halo. Of the 150 or so globulars known, the Messier catalogue contains 29.

Diffuse nebulae, clouds of dust and gas that lie in the plane of our galaxy, come in three varieties: *emission, dark, and reflection.* Emission nebulae are glowing (ionized) clouds of dust and gas that shine in the visible part of the spectrum. Dark nebulae are dense clouds of interstellar dust that obscure stars and parts of bright nebulae lying behind them, so we see them in silhouette. And reflection nebulae have no emissions of their own but scatter the incoming light from nearby stars. There are seven diffuse nebulae in the Messier catalogue.

Planetary nebulae are expanding shells of gas blown off dying stars about the same size of, or up to a few times larger than, our sun. The planetary shells last for about 50,000 years before they become too large and faint to be seen from earth. To astronomers looking through small telescopes a century ago (William Herschel, specifically), these nebulae looked like the planet Uranus (tiny round disks with a pale green tinge), and that's why they're called *planetary* nebulae. Of the 1,000 or so known planetary nebulae, the Messier catalogue contains four.

Finally, there are the galaxies. To see them we must peer through the Milky Way's dust, gas, and stars into the vastness of space, where other island universes are traveling on mysterious courses. In the grand scope of the known universe, it appears our young Milky Way spiral is in the minority. Most galaxies are older ellipticals – armless galactic nuclei. These ellipticals range in shape from round to very elongated. The Messier catalogue contains 9 ellipticals, 29 spirals, and 1 galaxy classified as *irregular.*

You can see a representation of nearly each one of these categories without optical aid. The following text spotlights some shining examples for each season. These objects are visible at different times of night at different times of the year. Use your planisphere to determine when you can see them and plan your nights accordingly.

SOME EASY-TO-FIND MESSIER OBJECTS
Winter. Assume it is 1 January at 9:00 p.m. Pick up your planisphere, go outside, and dial in that time and date. Turn the planisphere so that the southern horizon is down and look due south. About two-thirds up from the southern horizon you should see a tight grouping of six or more stars glittering in a dipper-shaped pattern. This is not the Little Dipper but the Pleiades, Messier 45 (M45), an open cluster. The cluster should be clearly marked on your star wheel. Now you can turn to the page dedicated to M45 in this book and start your adventure! Under dark skies, you might use binoculars or a telescope to search for hints of the faint reflection nebula that envelopes the Pleiades.

There's another winter splendor visible at the same time. Look high in the southeast, to the constellation Orion, conspicuous by its row of three equally bright stars that form Orion's belt. Less than two finger-widths south of (below) the belt is another chain of three stars, Orion's sword. The middle star in the sword will appear fuzzy to the naked eye. Turn a pair of binoculars to this patch of light and prepare to be impressed by the magnificent emission nebula (and star cluster) M42, the Orion Nebula. Your telescope will reveal another emission nebula, tiny M43, just north of it.

Spring. If you go outside on 1 April at 9:30 p.m. and face due south, you'll see the bright, blue-white star Regulus about halfway up the sky. Far to its west (right) are the twin stars Castor and Pollux (confirm these with your star wheel). Draw a line between Pollux and Regulus and look a little below the halfway mark, you should see a large fuzzy glow. This is M44, the Beehive, another open cluster celebrity.

The keystone (or square) of Hercules is rising at the same time in the northeast. Draw a line between the two western stars in the constellation's "keystone". About a third of the way along that line from the northernmost star is M13, the great Hercules globular cluster! If you do not live under dark skies, you may need binoculars to see it. Finding Messier objects with your binoculars is good practice before using your telescope.

Summer. On 1 August at 9:00 p.m. the curved tail of Scorpius is wonderfully placed for viewing. Look about a fist width above the southern horizon. Two finger-widths to the east (left) of the Scorpion's stinger (two stars very close together at the end of the tail) is a large hazy patch, M7, a fine open cluster. Binoculars resolve it into dozens of stars. Another open cluster, M6, lies about one binocular field to the upper right of M7.

The teapot of Sagittarius stands just to the east (left) of the Scorpion's tail. Under dark skies, the Milky Way band seems to steam out of the teapot's spout and waft across the night sky. Use your binoculars to scan that dense Milky Way region, because it contains many star clusters and nebulae, about a dozen of which are Messier objects. Here you will also find a network of dark nebulae weaving through the brilliant star clouds.

Fall. The great square of Pegasus dominates the high southern sky on 1 November at 9 p.m. Again, look due south, about two-thirds of the way up the sky and try to locate the four bright corner stars in the square. They mark the body of the mythical winged horse. Think big because the square averages about 15° (a fist and two fingers) on a side. Using the great square as a landmark, you can find a spiral galaxy with your unaided eye. The journey is a little more involved than the previous ones.

First locate the star that marks the northeast corner (upper left) of the great square. This star is called Alpheratz. Using your hand and fingers, look 15° to the northeast of Alpheratz, for a yellowish star of similar brightness. This is Mirach, in the constellation Andromeda. M31, the Great Andromeda Galaxy, is 10° (one fist-width) to the northwest (upper right). Under dark skies it should look like an oval glow to the unaided eye. Binoculars will show it much more clearly. Binoculars will also reveal its companion, M110, an elliptical galaxy, just northwest of it.

Only the tiny, dim glows of the planetary nebulae escape naked-eye detection. Actually, one planetary nebula, the Dumbbell (M27), is visible to the unaided eye, but only from the darkest sites and to those with exceptional eyesight or observing experience. As described later in this book, some planetaries can be detected with binoculars and a bit of perseverance.

Admittedly, not all the Messier objects are easy to locate and identify. But once you master navigating the sky and recognizing star patterns, you will be better prepared to track down the rest of them.

SOME TERMS

Turn to the beginning of chapter 4, where I describe M1, the Crab Supernova Remnant. There you will find a listing of the object's *NGC number, coordinates, magnitude,* and *dimensions.* What do these terms mean?

NGC is short for *New General Catalogue of Nebulae and Clusters of Stars.* Compiled by Danish astronomer J. L. E. Dreyer and published in 1888, this work lists 7,840 deep-sky objects in order of right ascension (described subsequently). The *NGC* includes most of the Messier objects. The Crab Nebula (M1) is entry number 1,952 in the *NGC;* thus M1 is also known as NGC 1952. Some objects have an *IC* number. The *Index Catalogues* were supplements to the *NGC.* For each object, I have included the description from the *NGC* or *IC,* as well as a translation of Messier's own descriptions. Together they will give you a historical perspective and a basis for comparison of your own observations.

The Crab Nebula's coordinates (5^h 34^m 30^s, $+22°$ $01'$) represent its *right ascension* and *declination,* respectively. Think of these terms as celestial longitude and latitude. If someone gave you the longitude and latitude of New York City, you could locate it on a map of the United States that contains a longitude and latitude grid. Likewise, you can locate the Crab Nebula on a star map using its celestial coordinates.

Turn to the inside back cover of this book where you will find a star map covering nearly the whole sky. Right ascension runs from right to left along the top and bottom of the map. Declination runs up and down the map's sides. Note that right ascension is written in hours (from 0 to 24 hours) and that the hours increase to the left (east). These lines of celestial longitude reflect the 24-hour rotation of the earth, which makes the entire celestial sphere *appear* to turn. Lines of declination mimic earth's latitude lines. They increase from 0° at the celestial equator (imagine the plane of earth's equator reaching out to the dome of the sky) to $+90°$ at the north celestial pole. They decrease from 0° to − 90° at the south celestial pole. To locate M1, trace the lines of right ascension eastward (to the left) until you find "5 hours," then continue until you're about halfway between "5 hours" and "6 hours." Next, move down until you reach a declination of + 22°. You should see M1 plotted there. A closer view of this region appears in the finder chart on page 40.

Unless otherwise noted, the coordinates used in this book are precise for "equinox 2000.0." The coordinate system is in constant change, because gravitational tugs by the sun, moon, and planets cause earth's axis to wobble like a top. It takes about 26,000 years for the axis to complete a wobble. Although this sounds like a long time (and it is), the gradual shift adds up, so every 50 years or so star charts are revised to incorporate this shift, or *precession,* of the coordinate system against the backdrop of stars. For this book, the coordinates given correspond exactly to the year 2000, hence equinox 2000.0.

Bright stars near the "M" objects are labeled with lower case Greek letters. This system of stellar nomenclature was introduced in 1603 by Bavarian astronomer Johann Bayer, who labeled stars in each constellation according to their brightness. The most prominent star was given the letter Alpha (α); the faintest became Omega (ω). The brightest star near M1, for instance, is Zeta (ζ), in the constellation Taurus. Astronomers condense it all by saying, "Zeta Tauri," which is the Greek letter followed by the Latin genitive of the constellation. There are exceptions, however, such as with the stars in the Big Dipper, which are labeled in order of right ascension, from west to east, not by brightness.

The Greek alphabet (lower case)

α	Alpha	ι	Iota	ρ	Rho
β	Beta	κ	Kappa	σ	Sigma
γ	Gamma	λ	Lambda	τ	Tau
δ	Delta	μ	Mu	θ	Upsilon
ε	Epsilon	ν	Nu	φ	Phi
ζ	Zeta	ξ	Xi	χ	Chi
η	Eta	o	Omicron	ψ	Psi
θ	Theta	π	Pi	ω	Omega

Other stars have number identifications. These are *Flamsteed numbers.* Like the Greek letters, a Flamsteed number precedes the Latin genitive of the constellation: 27 Tauri, for example. John Flamsteed was a prodigious eighteenth-century observer who dedicated 30 years of his life to measuring star positions, which he dutifully catalogued in his *Historia Coelestis Britannica* (1725) in order of right ascension. Interestingly, Flamsteed did not number the stars on his star charts; he used only the

Greek letter designations. The numbers were assigned later by Joseph Jerome de Lalande to stars in a French edition of Flamsteed's 1780 star catalogue.

Constellations and their latin genitive forms

Abbrev.	Constellation	Latin genitive	Abbrev.	Constellation	Latin genitive
And	Andromeda	Andromedae	Lac	Lacerta	Lacertae
Ant	Antlia	Antliae	Leo	Leo	Leonis
Aps	Apus	Apodis	LMi	Leo Minor	Leonis Minoris
Aqr	Aquarius	Aquarii	Lep	Lepus	Leporis
Aql	Aquila	Aquilae	Lib	Libra	Librae
Ara	Ara	Arae	Lup	Lupus	Lupi
Ari	Aries	Arietis	Lyn	Lynx	Lyncis
Aur	Auriga	Aurigae	Lyr	Lyra	Lyrae
Boö	Boötes	Boötis	Men	Mensa	Mensae
Cae	Caelum	Caeli	Mic	Microscopium	Microscopii
Cam	Camelopardalis	Camelopardalis	Mon	Monoceros	Monocerotis
Cnc	Cancer	Cancri	Mus	Musca	Muscae
CVn	Canes Venatici	Canum Venaticorum	Nor	Norma	Normae
CMa	Canis Major	Canis Majoris	Oct	Octans	Octantis
CMi	Canis Minor	Canis Minoris	Oph	Ophiuchus	Ophiuchi
Cap	Capricornus	Capricorni	Ori	Orion	Orionis
Car	Carina	Carinae	Pav	Pavo	Pavonis
Cas	Cassiopeia	Cassiopeiae	Peg	Pegasus	Pegasi
Cen	Centaurus	Centauri	Per	Perseus	Persei
Cep	Cepheus	Cephei	Phe	Phoenix	Phoenicis
Cet	Cetus	Ceti	Pic	Pictor	Pictoris
Cha	Chamaeleon	Chamaeleontis	Psc	Pisces	Piscium
Cir	Circinus	Circini	PsA	Piscis Austrinus	Piscis Austrini
Col	Columba	Columbae	Pup	Puppis	Puppis
Com	Coma Berinices	Comae Berenices	Pyx	Pyxis	Pyxidis
CrA	Corona Australis	Coronae Australis	Ret	Reticulum	Reticuli
CrB	Corona Borealis	Coronae Borealis	Sge	Sagitta	Sagittae
Crv	Corvus	Corvi	sgr	Sagittarius	Sagittarii
Crt	Crater	Crateris	Sco	Scorpius	Scorpii
Cru	Crux	Crucis	Scl	Sculptor	Sculptoris
Cyg	Cygnus	Cygni	Sct	Scutum	Scuti
Del	Delphinus	Delphini	Ser	Serpens	Serpentis
Dor	Dorado	Doradus	Sex	Sextans	Sextantis
Dra	Draco	Draconis	Tau	Taurus	Tauri
Equ	Equuleus	Equulei	Tel	Telescopium	Telescopii
Eri	Eridanus	Eridani	Tri	Triangulum	Trianguli
For	Fornax	Fornacis	TrA	Triangulum Australe	Trianguli Australis
Gem	Gemini	Geminorum	Tuc	Tucana	Tucanae
Gru	Grus	Gruis	UMa	Ursa Major	Ursae Majoris
Her	Hercules	Herculis	UMi	Ursa Minor	Urase Minoris
Hor	Horologium	Horologii	Vel	Vela	Velorum
Hya	Hydra	Hydrae	Vir	Virgo	Virginis
Hyi	Hydrus	Hydri	Vol	Volans	Volantis
Ind	Indus	Indi	Vul	Vulpecula	Vulpeculae

Magnitude refers to an object's apparent brightness. (There is an involved discussion of limiting magnitude – how faint one can see – in chapter 3.) Think of magnitude as "class." If something is "first class," it's great; anything else is lower. And that's how the Greek astronomer Hipparchus must have viewed the stars in the second century B.C. when he designated the brightest naked-eye stars as 1st magnitude and the faintest ones as 6th. Mathematically, a 1st-magnitude star is 2.512 times brighter than a 2nd-magnitude star, which is 2.512 times brighter than a 3rd-magnitude star, and so on. The math works out nicely so that a star of 1st magnitude is exactly 100 times brighter than a star of 6th magnitude (because 2.512 is the 5th root of 100). On the brighter end of the magnitude scale, the values soar into the negative numbers: the sun is magnitude –27; the full moon is magnitude –12.5; Venus can reach magnitude –4.9, and Sirius, the brightest star in the night sky, is magnitude –1.6.

Most astronomy books state, as a general rule, that the faintest star visible to the unaided eye is 6th magnitude; the faintest star visible with 7×50 binoculars is 9th magnitude, and 12th magnitude is the faintest you will see with a 4-inch telescope. However, these numbers are *very* conservative and should only be used when making generalities about average skies. For some reason these magnitude limits have become chiseled in stone and have misled beginners for decades! This is a very important point. Be sure to read the section on magnitude limits in chapter 3 for more discussion.

The *apparent size* of a deep-sky object is an angular measure of its dimensions against the celestial sphere. The units of angular measure are *degrees* (°), *arc minutes* ('), and *arc seconds* ("): $1° = \frac{1}{360}$ of a circle; $1' = \frac{1}{60}$ of a degree; and $1'' = \frac{1}{60}$ of an arc minute. The angular separation between the two pointer stars in the Big Dipper is 5°. Both the sun and moon are about ½° in diameter, or 30'. Most of the Messier objects have much smaller angular measures. The Crab Nebula, for example, is an irregularly shaped haze 6' long and 4' wide, or $6' \times 4'$, or about one-fifth the moon's diameter. (It is important to note, however, that although the moon appears large, this is an optical illusion. It's much smaller than the width of your little finger held at arm's length.)

The light of each Messier object is spread over a specific area of sky. If M1 shines at 8th magnitude, its light output is the same as an 8th-magnitude star. But its light also covers a larger area of sky than the star. Have you ever used a flashlight with an adjustable beam? Think of how bright the light appears when the beam is concentrated and how much weaker it looks when the beam is diffused. Likewise, a diffuse 8th-magnitude Messier object will appear dimmer than an 8th-magnitude star. This dimming effect is intensified under less-than-perfect sky conditions.

An object's magnitude alone, then, can be deceiving, especially with faint, diffuse objects like a galaxy. Therefore, to help provide you with a better measure of a galaxy's visibility, the data lists for galaxies in this book include a value for *surface brightness*. Think of it as dividing the object's magnitude by its area, though the math is more complicated than this.

Take the galaxy M65 in Leo, for example. It shines at magnitude 9.3 and occupies an area measuring 9'.8 × 2'.9. Its surface brightness, is 12.4, meaning that each arc minute of the galaxy shines roughly with the brightness of a 12.4-magnitude star.

That raises another key issue – *extinction*. Your latitude determines how far south on the celestial sphere you can see. For instance, if you live on the North Pole (+90°), you cannot see any Messier objects with declinations south of the equator (0°). Although all the Messier objects in this book are visible from northern latitudes, some will be closer to the horizon, where the atmosphere is densest. Generally, stars near the horizon appear dimmer by a few tenths of a magnitude than a star shining overhead. If the air is very polluted, a star's light could be diminished by one or more magnitudes! Because the Messier objects are large and diffuse, their visibility is greatly affected by such pollution.

STAR COLOR

The descriptions of the Messier objects, also spotlight bright, nearby stars, listing their magnitudes and spectral classification. These data were gleaned from *Sky Catalogue 2000.0*. The spectral classification reveals many physical characteristics of a star, including its surface temperature, size, and density.

Spectral classes are designated by the letters *O, B, A, F, G, K,* and *M,* which correspond to surface temperature. The hottest stars have the letter designation *O,* while the coolest stars have the designation *M.* You can remember this sequence by the mnemonic "*O*h, *Be A F*ine *G*irl (Guy), *K*iss *M*e." A star's apparent color is directly related to the temperature of the gas at its surface and to its surface area. Small blue stars, for example, are very hot (some 40,000 K [Kelvin]), while red giant stars are relatively cool (about 3,000 K). (The Kelvin scale begins at absolute zero – about −273.16 °C [degrees Celsius], the coldest temperature that can be approached. Water freezes at 273 K and boils at 373 K.)

Each spectral type is divided further into 10 subclasses denoted by the numbers 0 through 9. The higher the number the cooler the star. Our sun, for example is classified as a *G*2 star, which is slightly cooler than a *G*1 star and slightly hotter than a *G*3 star.

Finally, a star's *luminosity* class is indicated by a roman numeral from I to VI. Supergiant stars are I; bright giants, II; giants, III; subgiants, IV; main sequence stars/dwarfs, V; and subdwarfs, VI. Thus, Zeta (ζ) Tauri, designated as a *B*2 IV star, is a hot, blue, subgiant star.

The following table lists the spectral types, apparent color, approximate surface temperature, and an example of a familiar member of each class.

Spectral type	Apparent color	Surface temperature (K)	Star
O	blue	25,000–40,000	Zeta Orionis (O9)
B	blue	1,000–25,000	Spica (B1)
A	blue to white	7,500–11,000	Vega (A0)
F	white	6,000–7,500	Polaris (F8)
G	white to yellow	5,000–6,000	Sun (G2)
K	orange to red	3,500–5,000	Arcturus (K2)
M	red	3,000–3,500	Antares (M1)

FINDING YOUR TARGET

When planning your night's observing, select an object you would like to see, read the translated Messier catalogue and *NGC* descriptions of its telescopic appearance, and examine the accompanying photograph and drawing. (Place a book mark on the page(s) for that object, because you will refer to it often in the field.) Next, scan the all-sky map in the back of the book to find the object's general location, the constellation it is in, and a nearby bright star or two to use as guideposts. (The brighter stars and other reference stars are labeled with their identifying Greek letters or Flamsteed numbers.) Now turn to the smaller-scale finder chart in the section of the book describing that object to zero in on your destination. Some of the finder charts show stars to 6th magnitude, others to 7th or 8th magnitude, depending on the chart's scale. Once you locate and identify your target, use your red flashlight to read the more extensive notes, which will steer you to several visual attractions in and around that particular object.

One thing you can start doing now is to practice aiming your telescope on one of the bright seasonal Messier objects mentioned earlier. Do so until it becomes second nature, because you cannot expect to hit your celestial target if your telescope and finder scope are not carefully aligned. Nothing is more frustrating than hunting for a faint Messier object when you're starting with the wrong guide star. It is amazing how similar the star fields appear when you are looking through a telescope. Take the time now to practice aiming. You can also do this during the day using a distant object on the horizon.

Pointing with accuracy is impossible unless you know how to scale what you see through the eyepiece with what you see on the finder chart. The maps show more sky than you can see through your telescope, and your telescope will reveal many more stars than are plotted on the maps. Therefore you have to determine what area of sky you can see with your low-, medium-, and high-power eyepieces. To determine the *field diameter*, simply choose any star near the celestial equator (such as one of the stars in Orion's Belt) and time how long it takes for the star to drift across the entire field. When multiplied by 15, this number will give you the *field diameter* in minutes and seconds of arc. For instance, if it takes five minutes for the star to drift across, then your field diameter is 75′, or 1¼°. Most telescopes at low power will show at least 1° of sky.

There are electronic devices now that can effortlessly guide you (or your telescope) automatically to thousands of celestial objects, virtually eliminating the time spent searching for them. But you should save that wizardry for star parties, when you are showing the sky to friends, or on nights when time is of the essence. For now, think in terms of "exploring the sky." Besides, you do not want to be so dependent on machinery that you are helpless without it.

How do you find the fainter Messier objects? First, use the wide-field map in the back of the book to get a ballpark location for the "M" object you want, noting its position among the brighter stars. Now look at the detailed finder chart on the object page and see which naked-eye stars make a pattern with the object – simple patterns like a triangle, a straight line, or a square. Next, find those stars in the sky. (I always use binoculars to confirm the star field.) and place the cross hairs of the finder scope where the Messier object fits into the geometrical pattern, even if you cannot see the object. When using a finder scope, I keep both eyes open; one eye is focused on the cross hairs, and the other eye is looking at the sky. If your telescope is properly aligned, and if you have practiced your aiming, you should be very near, if not right on, the object. Use the lowest magnification (the one with the widest field of view) and look for a faint, diffuse patch of light.

If the object is not immediately recognizable, try to identify the stars in the telescope field with those on the finder chart. You might have to hop from star to star to reach your destination. To do this, a knowledge of field diameter and sky direction is invaluable. If you get confused, return to a bright, easy-to-find Messier object, like the Pleiades or the Orion Nebula, and practice moving your telescope up and down, left and right, to determine the field's diameter and orientation.

Note that the finder charts in this book are all oriented with north up and west to the right. The unnerving part about field orientation is that a telescope inverts the image, so south is at the top and east is to the right. To match the view through your finder scope, you will have to rotate the book. Furthermore, if your telescope is on an equatorial mount, the field will often be rotated with south to the upper right or upper left. Nudge the telescope tube *toward Polaris* and notice where stars enter the field – that is north. For telescopes on an altazimuth mount, as mine is, just tilt the star map until it matches your view.

If your telescope's finder scope is equipped with a diagonal, remove it and replace it with a regular eyepiece (or invest in a new finder); also avoid using a diagonal with your telescope when you are using a star chart, unless you want to retrace every star chart so that it represents a mirrored view. Finally, if you are not having any success in your search, take a break, then start over. It takes practice to hit your mark.

OBSERVING TIPS

Have you ever left a theater in the middle of the movie to get some popcorn? You leave confident knowing you can find the way back to your seat, because you can see fairly well in the low light level. You enter the brightly lit foyer, get the popcorn, and return to the theater. Suddenly, you can't see a thing except the screen. You walk down the aisle with arms extended like Frankenstein's monster, groping for the right seat. Several minutes later your night vision returns and everything appears normal. The same thing happens every time you step outside from a lighted house and stand under the stars.

The retinas of our eyes are packed with light-sensitive neural receptors called *rod* cells and *cone* cells. The cones are less sensitive to light than the rods and therefore work best in daylight. When we go from a very bright to a very dark environment, our cones essentially shut down, but our night-sensitive rods do not kick in instantly. So we become temporarily blinded. The visual pigment *rhodopsin* in the rods bleaches out in intense light and takes about 30 minutes to regenerate. It takes about 30 minutes for our eyes to fully *dark-adapt*. You can start seeing things a lot sooner than that, but don't expect to discern really faint details in a Messier object until you are fully dark adapted. For instance, tests have shown that eyes dark-adapted for 30 minutes are six times more sensitive to light than eyes dark-adapted for 15 minutes. My experience in the field has been that this time varies with the individual. I don't know if there is any way to validate what I do to hasten the regeneration of rhodopsin in my eyes, but the following *seems* to work. I pull my coat, jacket, or sweater over my head and stare intently into the darkness for a minute or two. Other times I pick the darkest spot around and just stare at it, occasionally squinting real hard, until more and more detail becomes visible.

Rod cells are also peculiar in that they do not lie directly in the center of vision (that is where the cones are), but are in the perimeter. Faint celestial objects will be best seen using *averted*, or peripheral, vision. In other words, do not center a faint galaxy in the eyepiece and stare directly at it – the galaxy will probably vanish, because you are trying to see it with your day-sensitive cone cells. Look off to one side a little but focus your *attention* on the object.

I tend to favor a particular spot in my right eye when looking for faint details. To see a knot in a galaxy's faint spiral arm, for example I have a tendency to position the galaxy to the upper left of the field, in the direction of my temple. So, if the galactic knot is at the 11 o'clock position, I direct my line of vision to the 4 o'clock position. Since our eyes interact with specific parts of the brain, this action might reveal a subconscious knowledge of the most sensitive region in my field of rods. Perhaps a bunch of rods work together at that spot to enhance my vision. In any case, I've developed an awareness of how my own mind and eyes react to detecting faint objects. Try to discover that magic spot in your own eye or eyes (if there really is one), or just find the most comfortable position for looking. Your magic spot may be totally different from mine.

The late Carolyn Hurless, a prolific observer of variable stars, shared the following observing secret with me when I was about 16 years old. She jokingly called it "heavy breathing," and said it was a tactic that the late variable-star observer and comet discoverer Leslie Peltier employed when trying to detect very faint variable stars from his observatory in Ohio. The trick is to hyperventilate, taking several very deep breaths (actually through your nose with your mouth closed) before you put your eye to the eyepiece. This sends a fresh supply of oxygen to the brain and eyes and increases your alertness. Take a few minutes to scan the field with this fresh oxygen supply, then begin inhaling and exhaling slowly through your nose. (Actually I now do this through puckered lips. Luckily I'm in the dark!) Increase the frequency (so the breaths are shorter and deeper) as you zero in on a target. This reminds me of the way a bat hunts, using echolocation. It sends out pulses of sound at a certain frequency until it finds a meal, then it zeros in by increasing the frequency of the pulses. You have to be careful, however, not to overdo the heavy breathing. On a few occasions, I have nearly passed out trying to see faint stars!

Hyperventilating is great for detecting stellarlike objects and tiny faint patches of nebulosity. But very large and diffuse objects present a different challenge. Take something like the Andromeda Galaxy, Orion Nebula, or nebulosity in the Pleiades. These are very extensive glows. The question is, how do you determine how far the object extends before it fades into the bright starry background? The nineteenth-century Harvard astronomer George Bond would place the object well out of the field of view of the great Harvard 15-inch refractor and let the object drift back into view. When he suspected seeing a change in the sky background, he would make a note. And this is essentially what I have done for the larger Messier objects. But you can also manually move the telescope far afield and gradually bring the object back in (without waiting for the earth to rotate) and achieve the same result.

If I am not sure whether some faint detail is real or not, I will jiggle the telescope ever so slightly to set the sky in motion. The eye is better at detecting tiny moving objects than tiny stationary ones. For example, it is easier to detect a very faint satellite moving among a myriad of stars than it is to detect an equally faint stationary nova.

Although the effects of diet and other factors on visual observing are still being studied, being in good physical condition helps night vision and seems to quicken dark adaptation. It is best to be well rested; I can see one full magnitude fainter when I'm rested than when fatigued. On the moonless nights during which I made the observations for this book, I exercised regularly; avoided alcohol; wore warm, comfortable clothing; and observed from a relaxed position.

Finally, the most important factor in detecting faint details is the amount of time you spend observing. Don't rush through the Messier objects like you are at a pie-eating contest (see my comments on "Messier marathons" in appendix B), but rather, savor each one. The more time you spend looking at a particular object over the course of a night or several nights, the more detail you will see. That is how I approached the observations in this book. The following drawings show how, over time, I could see more detail in the galaxy M101 in Ursa Major.

For most of the Messier objects, I observed each for an average of six or more hours over three nights. Some of the more complex objects were observed for three hours per night for several nights. Don't expect to see all the detail in my drawings with just a glance. Challenge yourself to spend the time to really study these objects, which are some of the most splendid deep-sky wonders visible from our unique perspective in space and to strive to see even *more* detail than I have shown. Once you have located a Messier object, let's say a galaxy, spend about a half hour just enjoying it. Next, try to sketch as much detail as you can without referring to the book. (Even if you can't draw, make a sketch; you will be surprised at how your artistic talents will increase by trying.) Then sit back for a moment, take a sip of hot tea, relax your eyes, examine the photograph and drawing, and compare your sketch with the photograph. Before you return to the eyepiece, read the descriptive information on that page and pick out a specific detail you want to find; maybe this particular galaxy has a faint outer arm, which you overlooked. When you return to the eyepiece, you now focus your attention only on trying to see that particular detail. If you do this for each feature within that particular object, using different magnifications, your drawing will ultimately come together into a coherent whole! By training your eye in this way, you will be able to pick up these subtle features more easily on subsequent observations of the object.

3 The making of this book

This book is your companion under the stars. Similar to a field guide to birds, insects, or flowers, this book will help you locate and identify the Messier objects – the most famous deep-sky splendors in the universe. It contains, among other things, finder charts to help you locate them, photographs, pencil drawings of their telescopic appearance, detailed descriptions, and descriptions of other interesting celestial sights near the Messier objects. I have also included a lineup of some of my favorite non-Messier objects.

Each Messier object section opens with a photograph and some essential data, including the object's coordinates, magnitude, apparent size, and distance. Descriptions of the object from a translation of Charles Messier's original catalogue and from the *New General Catalogue of Nebulae and Clusters of Stars* (1888) or the two subsequent *Index Catalogues* (1895 and 1908) follow. These data and descriptions should help you realize what to expect when you first glance at a Messier object through the eyepiece.

The objects are ordered as Messier catalogued them. However, you can view them in any order you wish, as long as a particular object is above the horizon when you plan to observe it. Use a star wheel to determine which constellations and, therefore, which Messier objects are visible on the date and time of night you want to look.

THE TELESCOPE

This book was conceived at the May 1994 Texas Star Party, where I spent several nights observing the Whirlpool Galaxy (M51) with a Tele Vue Genesis 4-inch refractor. Until that time I had spent most of my life observing the moon and planets with permanently mounted, long-focus, observatory-class refractors with old, high-quality glass. Any views of the deep sky were confined to high powers and narrow fields. For instance, the lowest magnification I normally used with the 9-inch f/12 Clark refractor at Harvard College Observatory was $137\times$, which offered me a $\frac{1}{2}°$ field of view. Only one-ninth of the Andromeda Galaxy (M31) would fit in that field! And when I used the 18-inch f/16 Clark refractor at Amherst College, a typical low power was $365\times$ and the field of view was even smaller. Even as a young Messier-object hunter in the 1960s, I used long-focus reflectors and refractors and never saw the objects in a field greater than 1°. The Genesis refractor, when coupled with a 22-mm (2-inch) Nagler eyepiece, will show the Andromeda Galaxy at 23× in a field of view close to 3°! Nearly the entire galaxy and its companions fill the field. Furthermore, changing eyepieces and adding a Barlow lens allows me to study the galaxy's nuclear region with 10 times that magnification.

When it came time to decide which telescope to use for this project, I had no hesitaion in choosing the 4-inch f/5 (500-mm) Genesis refractor. The unobstructed optics in the Genesis are of unquestionable quality. Al Nagler (formerly a NASA optical engineer who designed the wide-field optics for the Apollo Lunar Landing Simulator) created a special four-element optical primary for the Genesis refractor. When coupled with his Tele Vue eyepieces (which have an equally revolutionary optical design), the Genesis transcends the traditional limitations of both the long-focus and rich-field refractors by combining the best qualities of both.

For this study I used only two eyepieces: a 22-mm Panoptic (adaptable 1¼-inch and 2-inch) and a 7-mm Nagler (1¼-inch); these provided magnifications of 23× and 72×, respectively. A 1.8× Barlow lens with the 7-mm eyepiece gave me a "high" power of 130×, which is a magnification of about 32× per inch of aperture. As a rule, 50× per inch of aperture is considered the maximum useful limit under ideal atmospheric conditions. Although the Genesis can perform beyond that theoretical limit (especially under very dark and stable skies at high altitudes), I restricted myself to the above magnifications, because the vast majority of us constantly deal with less-than-perfect atmospheric conditions.

I did not use any filters when making the observations for this book. Observers who live under light-polluted skies, however, should consider using light-pollution or skyglow-reduction filters to sharpen the contrast between the diffuse Messier objects and the background sky. Hydrogen-beta and oxygen-III filters work wonders on emission nebulae, but be aware that they also dim starlight by about one magnitude.

The Genesis is cradled in the altazimuth yoke of a Gibraltar mount. This heavy, ash-wood tripod certainly lives up to its name and holds the telescope rock steady. Sometimes when I am out observing on the summit of Kilauea volcano, the winds will suddenly gust to gale force, yet the telescope hardly jiggles. The mount has also performed quite heroically during several minor earthquakes (a frequent occurrence when you live on an active volcano!); I never lost my field of view during these events, and sometimes the small volcanic quivers helped me to confirm the existence of some faint details in a nebula.

I should add that no one twisted my arm, dangled a carrot, or offered me the key to Manhattan Island to use this particular telescope for this project. The Genesis and I met serendipitously under dark Texas skies, where I realized its quality and potential. The rest is history. Naturally, you can enjoy the Messier objects through any telescope, large or small. And most of the objects are even visible in 7 × 35 binoculars.

OBSERVING LOCATION

After the 1994 Texas Star Party, I returned home to Boston, Massachusetts, where city lights wash out all but the brightest stars from the night sky. Unquestionably, if this project were to go forward, I would have to travel to complete the book. My plan was to observe from the darker skies of western Massachusetts on weekends and augment these observations with any I could make at various star parties. By a twist of fate, in the fall of 1994 my wife, Donna, was offered a job on the Big Island of Hawaii, and I followed her there that December.

A better stage could not have been set for performing the observations for this book. We purchased a house in Volcano, Hawaii, at an altitude of 3,600 feet. It is some 3 miles east of the summit of Kilauea volcano, which rises 4,200 feet above sea level. The nearest streetlight to our house is 1 mile to the north. Beyond that some 300,000 acres of Hawaii Volcanoes National Park and reserved forest border our subdivision to the north, south, and west. Hilo, the nearest city to the east, is about 45 miles away. The massive, swollen back of Mauna Loa volcano, which rises nearly 14,000-feet above sea level, blocks our view of Kona, 100 miles to the west. All around us for 3,000 miles is the vast Pacific Ocean. Adding to the visual pleasure, the entire island has a lighting ordinance to help keep the sky dark for astronomers at the 14,000-foot-high Mauna Kea observatory to the north.

Here in Volcano, on clear moonless nights, the Milky Way is bright enough to cast shadows, and Venus can "pollute" the sky with its brilliance. The bright haystack of the *zodiacal light*, the dim, dusty alley of the *zodiacal band*, and the much fainter elliptical glow of the *gegenschein* are all visible from our front yard. These large, faint features of the night – created by sunlight reflecting off dust-size particles in the plane of our solar system – are the hallmarks of a truly dark observing site. In their book, *Observing Handbook and Catalogue of Deep-Sky Objects*, Brian Skiff of Lowell Observatory and Christian Luginbuhl write, "Frequently the sole requirement for the visibility of a notoriously difficult nebula is not a large telescope or a special eyepiece, but a truly dark sky." Although none of the Messier objects are notoriously faint or difficult, the exotic details of each – a spiral arm, a dark lane, a tenuous wisp of nebulosity – can be extinguished in all but the darkest of skies.

To push the limits of vision in deep-sky exploration (which I have tried to do for the observations in this book), you need to know the quality of your observing site. The following formula created by Tim Hunter, president of the International Dark-Sky Association, will help you rate your site:

$$OSI = (T + W + A + S + V) / C,$$

where OSI is the observing site index, T is the transparency (limiting magnitude), W is the weather (percentage of clear nights per year), A is the altitude (in thousands of feet), S is the seeing (1 to 10; $1 =$ worst), V is the visibility (percentage of unobstructed sky), and C is the convenience factor ($0 =$ a very inconvenient site, costs a lot to get there; $1 =$ a very convenient site).

The OSI of The Ka'u desert area of Volcano, where I do most of my observing, for example, would work out to be the following:

$$OSI = (8.0 + 0.75 + 4,000 + 8 + 0.9) / 90 = 45.$$

In this formula, the world's finest observing sites would rate 50 or greater. As fate would have it, the observations for this book were made during an exceptional period of atmospheric clarity. Since 1985, Kevin Krisciunas of the Joint Astronomy Center in Hawaii has measured the natural brightness of the night sky over Mauna Kea. The brightness varies during cycles of sunspot maxima and minima; heightened solar activity energizes earth's upper atmosphere, causing it to glow feebly but perceptibly. His data, plotted in the accompanying figure, show a significant decrease in sky brightness in 1995. In fact, in mid-1995, the sky over Mauna Kea was the darkest it had been in the last 10 years. The observations in this book were made over the seven-month period between December 1994 and June 1995. The sun experienced sunspot maximum in 1986 and probably reached minimum in 1995.

Yearly Averages of V-band Night-Sky Brightness

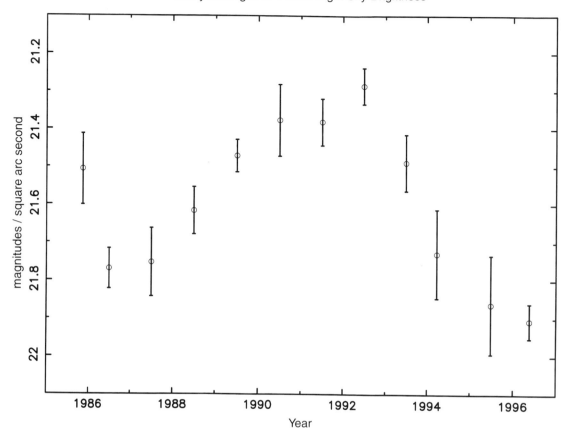

LIMITING MAGNITUDES

That truly great skies are truly dark skies is a myth. What is really meant by dark skies is that the atmosphere is so free of man-made pollutants and nighttime lighting that multitudes of fainter stars can be seen. The more starlight you see, the *brighter* the night sky appears. From the darkest sites, the zodiacal light and band, the gegenschein, and the Milky Way add illumination to the celestial backdrop. (This dark-sky myth is deeply rooted. Sometimes, visitors atop Mauna Kea – the world's best observing location, where magnitude 8.5 stars are within grasp of the naked eye – complain that the site is substandard, because the sky does not look *dark!*)

Limiting magnitude is itself a very good indicator of a sky's quality. But another myth can blind us from determining that limit accurately. Many popular amateur astronomy books state that 6th magnitude is the limit of naked-eye vision and that if you have good eyes and a good site you can probably see about 0.5 magnitude fainter. Magnitude limits are also attributed to telescopes of given apertures: a 4½-inch aperture will

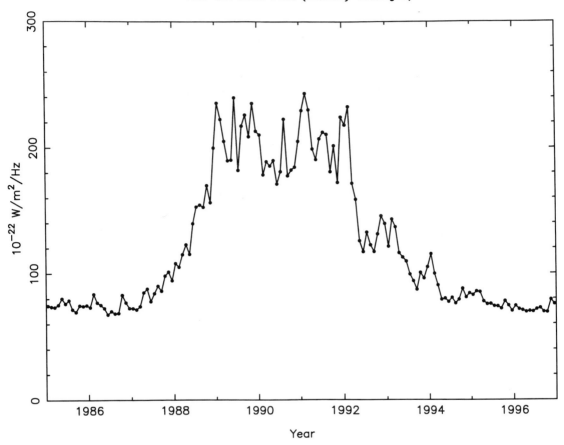

Year

show stars to magnitude 12; an 8-inch will reveal stars to magnitude 14. What these general rules fail to consider is the human contribution in seeing faint objects. Several years ago I researched the origins of the magnitude formula. Most profound is the classic work by the nineteenth-century English astronomer Rev. N. R. Pogson (*Monthly Notices* of the Royal Astronomical Society, Vol. XVII, 1857). His words, which reveal the individuality of limiting magnitudes, seem to have been forgotten:

> I selected [the ratio] 2.512 for the convenience of calculation. . . . If then any observer will determine for *himself* the smallest of Argelander's magnitudes just discernible by fits, on a fine moonless night, with an aperture of one inch, and call this quantity L, or the limit of vision for one inch, the limit l, for any other aperture, will be given by the simple formula, $l = L + 5 \times \log$ aperture. Numerous comparisons, made with various telescopes and powers, at different seasons of the year, have furnished me with the value $L = 9.2$ for my

own sight, which is, I believe, a *very average one*, and therefore suitable for such a determination.

All emphases are mine. Pogson created a formula that worked for him. He encouraged others to discover their *own* limits with this formula. Pogson clearly states that his eyesight was average, and he was keenly aware that limiting magnitude varies with aperture, season, and atmospheric conditions. Also, consider that the glass he used for his observations was inferior to that employed by today's amateurs. A better formula is the "brat" equation

$$N = brAtf(m)C,$$

where N is the number of photons available per unit time to the eye; b is the most effective bandpass for the human eye (100 nm); r is the transmission through the atmosphere, optics, and reflective coatings; A is the light-collecting area of the aperture (πr^2); t is the eye's storage time for collecting photons (0.1 second); $f(m)$ is the decrease in stellar magnitude, m, being discussed ($f(m) = 2.512^{-m}$); and C is the value of incident stellar radiation of a zero-magnitude standard star beyond the earth's atmosphere. $C = 10,000$ photons/second/nanometer per square centimeter.

The origin and significance of this formula is explained in more detail in an article entitled "Some Thoughts on Limiting Visual Stellar and Cometary Magnitudes With Various Apertures," by Daniel W. E. Green, in the *International Comet Quarterly* (April 1985). The equation demonstrates that under ideal conditions, the naked eye can, theoretically, detect enough photons to see a 9th-magnitude star! It is also not impossible to see stars fainter than 15th magnitude with a 6-inch reflector, as the keen-eyed variable star observer E. H. Mayer has done from his dark-sky site in Ohio.

In 1901 Heber Curtis of Lick Observatory wrote a memoir "On the Limits of Unaided Vision" (Lick Observatory *Bulletin* Number 38). He began with the following comment: "It is generally stated that stars of the sixth magnitude are as faint as can readily be seen by the unaided eye, though it is well known that under conditions of exceptional clearness favorably placed stars from a half to a whole magnitude fainter can be made out." Curtis proved this in a visual experiment conducted at the observatory in which he detected with the unaided eye stars of magnitude 8.2 without difficulty and stars of magnitude 8.3 with difficulty. He also glimpsed a star of magnitude 8.5. (Brian Skiff notes that Curtis's data are not on the new standard magnitude system. It might be interesting, Skiff suggests, to refer to the original paper and do this study again with new magnitudes for the stars Curtis saw.) Regardless, Curtis's observations support my own limiting-magnitude studies at the 9,000-foot level of Mauna Kea, where I consistently detected stars as faint as 8.4 magnitude with the unaided eye. In another study performed at the 1994 Texas Star Party, Florida amateur astronomer Jeannie Clark reported seeing a 7.9-magnitude star with her unaided eye. (For another discussion on the power of vision, see "Telescopic Limiting Magnitudes," by Bradley E. Schaefer, a pioneer in the study of human perception in astronomy, which appeared in the *Publications* of the Astronomical Society of the Pacific, volume 102, February 1990.)

Although "conventional wisdom" says you need an 8-inch telescope to resolve 14th-magnitude stars in globular clusters, in fact it is possible to reach that magnitude with a *4-inch* telescope. The same misconceptions apply to naked-eye and binocular magnitude limits as well. Of course, there are factors that influence the limit of our vision, the most debilitating of which is light pollution – that ugly glow over our cities and suburbs that robs us of our views of the stars. Light pollution is amateur astronomy's greatest nemesis. If writing this book has done anything for me, it has made me appreciate how beautiful the night really can be. I fear that if we do not act now to preserve the few untainted sites we have left for amateur astronomy, night itself may become a myth.

At the 1994 Texas Star Party Brian Skiff and I used a 7-inch refractor to do some visual tests of limiting magnitudes. When we turned it on the globular cluster M3 and compared what we saw to a magnitude sequence published in the *Observing Handbook and Catalogue of Deep-Sky Objects* (see photo opposite), we independently recorded a 15.1-magnitude star and concluded we could have gone fainter. That same magnitude sequence is reproduced here, to help you determine your own limits. And do not forget that using averted vision – looking slightly off to the side of the object of interest – can be very helpful when viewing faint objects. Skiff concludes from naked-eye tests that he can see *three magnitudes fainter* with optimally averted vision than he can with direct vision when his eyes are fully dark-adapted.

Altitude also has a significant effect on how faint you can see. The higher you are the fainter you can see, up to a point. I gain roughly ½ to 1 magnitude for every 3,000 feet above sea level. The improvement levels off at about 9,000 feet, for two reasons. First, you're above the densest layers of the earth's atmosphere (which greatly affect the visual limit), and second, the relative lack of oxygen in the air at that altitude means less oxygen gets to the eyes, thus adversely affecting vision.

MAGNITUDE ESTIMATES AND OTHER DATA

Estimating the brightness of diffuse objects is not easy. The traditional method, first employed by J. B. Sidgwick, is to compare the size and brightness of the diffuse object with a selection of similarly bright stars that have been racked out of focus until they appear the same size as the

diffuse object. In another method, both the star and the diffuse object are simultaneously defocused and their brightnesses and sizes compared. The results of these methods can be quite accurate, but it has been my experience, particularly in comet studies, that they are inadequate for objects with strong central condensations surrounded by faint extended halos. Many Messier objects, especially the galaxies and globular clusters, fall into this troublesome category.

Simple in-and-out focusing methods lead one to underestimate the intensity of these awkwardly diffuse objects, because the contribution of light from the extended halo is either negated (in the first method) or lost (in the second).

Twenty years ago, I devised my own method for estimating the magnitude of diffuse objects, which factors in the light loss of this outer envelope. This is accomplished by racking the diffuse object out of focus until it is a uniform glow. I then defocus my eyes to select candidate comparison stars. The selected stars are racked well out of focus, and I compare the glow from each stellar disk to that of the out-of-focus diffuse object. I then move back and forth between the diffuse object and the comparison stars until a reasonable match is obtained in both intensity and size. The entire procedure can take more than an hour! Try this method and see if you come up with magnitudes similar to those in this book.

Charles Morris, a well-known comet observer, has described a similar method in which the entire diffuse glow of the defocused comet is compared to a star image racked out to the same size (meaning the star is defocused even more than the comet). His method is published in the *International Comet Quarterly*, volume 2, 1980.

Nearly all my magnitude estimates for the Messier objects were made with the unaided eye or with binoculars. I included them in the list of data for each object only if I found a marked disagreement with published values. Otherwise, the magnitude estimates and other object data provided are from the following sources.

Stellar magnitudes, spectra, and double star information were gleaned from *Sky Catalogue 2000.0*, volumes 1 and 2 (Sky Publishing Corp., 1982 and 1985, respectively). Magnitudes and dimensions of diffuse nebulae are from Luginbuhl and Skiff's *Observing Handbook and Catalogue of Deep-Sky Objects* (Cambridge University Press, 1990); planetary nebula data were extracted from the *Catalogue of the Central Stars of True and Possible Planetary Nebulae*, by Acker and Gliezes (Observatory of Strasbourg, France, 1982). The sizes, magnitudes, distances, and star counts in open clusters are from *Star Clusters* (Willmann-Bell, 1996), by Steven Hynes and Brent A. Archinal. Accurately determining the number of stars in an open cluster, or its diameter, is virtually impossible because of the difficulty in discerning where the cluster proper ends and the background stars begin. Also, counts are sometimes based on measured, rather than only counted, stars. Therefore, published counts tend to vary.

Globular cluster magnitudes are from "Integrated Photometric Properties of Globular Clusters," an article by C. J. Peterson published in the Astronomical Society of the Pacific's conference series (volume 50, 1993), and their sizes and distances are from Skiff's "Observational Data for Galactic Globular Clusters," in the January 1995 issue of the Webb Society's *Quarterly Journal*. As for open clusters, determining the number of stars in globular clusters is an inexact science. The star counts given for globulars in this book are probably on the high side.

Magnitudes and diameters of galaxies are from the *NASA Extragalactic Database*, which was kindly provided to me in part by Barbara Wilson and Kevin Krisciunas. Galaxy velocities and distances are from the *Nearby Galaxies Catalog*, by R. Brent Tully (Cambridge University Press, 1988). Astronomers continue to debate galaxy distances.

Defensible estimates for the distance of the Virgo Cluster of galaxies, for example, range from 49 million to 72 million light years. The galaxy distances in Tully's work (and in this book) assume a Hubble constant of 75 kilometers/second/megaparsec and velocity perturbations in the vicinity of the Virgo Cluster. This model also assumes that the galaxy is retarded by 300 kilometers/second from universal expansion by the mass of the Virgo Cluster – which is some 55 million light years distant.

The physical diameters of the galaxies were calculated using the following formula:

$$\text{diameter} = 0.292 \times D \times R,$$

where D is the object's diameter in arc minutes and R is the object's distance in megaparsecs. The physical diameters of all the other Messier objects were calculated using a similar formula:

$$\text{diameter} = 0.000292 \times D \times R,$$

where D is the object's apparent diameter in arc minutes and R is the object's distance in light years.

Much of the data in this book differs from those appearing in older but popular references such as *Burnham's Celestial Handbook, The Messier Album,* and *Messier's Nebulae and Star Clusters.* This book contains the most up-to-date astronomical data for the Messier objects of any book in the popular literature.

THE PHOTOGRAPHS

All the photographs of the Messier objects in this work appear in black and white. Through the telescope, some stars, clusters, and nebulae do display subtle hues (though some astronomers would argue otherwise). But for the most part, the objects appear as grayish white hazes or clusters of stars projected against a black background. Color photographs, while often beautiful, may mislead beginners into thinking they can expect to see vivid reds, greens, and blues; such expectations would only lead to disappointment when the real thing is seen. Because my mission is to excite you about what you *can* see through a telescope, I decided to stick with black and white images only.

The photographs are meant not only to inspire you to look at these objects but also to help you confirm your visual impressions of them – especially when you *think* you're seeing an elusive detail in a nebula, or a pattern of faint stars in a cluster. For example, if you study the globular cluster M13 at high magnification, you might suspect several dark lanes that slice across the stars near the cluster's center. By orienting the photograph to match your visual impression, you will discover that the dark lanes really do exist. These features were first seen by Lord Rosse with his

72-inch speculum-mirror telescope in the 1800s, but they're within reach of moderate-size amateur telescopes today.

The photographic appearance of stars and nebulosity depends on the length of the exposure, the sensitivity of the film, and how the film is developed. Many of the photographs in this book show more stars and nebulosity than you can expect to see with your telescope. Yet there is also a frequent and unfortunate tradeoff in deep-sky photography. To expose the faint outer arms of a spiral galaxy, the photographer must overexpose the brighter core or nuclear region. Therefore, it is possible you will see more detail in these areas through the eyepiece than is apparent in the photographs. The same tradeoff applies to highly condensed globular clusters. In photographs, the tightly packed stars in a cluster's center often blend into an unresolvable mass.

For ease of orientation, all photographs in this book have south up and east to the right. This will match the view in a simple inverting tele-scope without a diagonal.

Of course, no photograph can record emotion. When I finally pick out a faint blur in the sky and realize that it is a galaxy 40 million light years away, well, the visual image might not be all that impressive, but I am in awe of it nevertheless.

THE DRAWINGS

All the drawings in this book are composites, based on several observa-tions with magnifications of 23×, 72×, and 130×. I spent a minimum of six hours on each object over three nights. The observations were made only on the clearest, moonless nights, at altitudes no lower than 3,500 feet. The drawings represent what good observers from dark, sea-level sites might expect to see with an 8- to 10-inch Schmidt–Cassegrain, a popular backyard telescope.

I did not try to accurately position every resolvable star in a globular cluster. I did, however, try to plot, to the best of my ability, the locations of any particularly bright members, especially if they would help you orient the drawing to the photograph for comparison. Otherwise, the drawings of the globular and open clusters reveal the patterns of major star streams, dark lanes, and irregularities in shape. Diffuse objects, such as the Orion Nebula, were extremely challenging and required longer dedicated efforts. I treated the Orion Nebula, for example, as if it were a dozen indi-vidual objects, and focused on one area at a time over the course of several weeks. (For a revealing look at how increased time spent behind the eye-piece can enhance the amount of detail you can see, refer to the sequence of drawings of the galaxy M101 on page 24.)

Although I strived for accuracy in the drawings, my renditions are not perfect, and my interpretations of what I see may be different from those of others. For instance, my eye and mind might work together to follow a particular pattern of stars in a globular cluster – say, a counter-clockwise spiral of stars – but other observers might see a perfectly reasonable *clockwise* spiral of stars in the same region.

For some objects, open clusters in particular, I took the liberty of drawing the whimsical creatures (e.g., bats, alligators, fireflies) that I visualized in the patterns of the stars during moments of fancy. An example is M41, whose stars form an outline of a fruit bat reaching for a bite to eat. I have also highlighted certain geometrical patterns in the drawings for emphasis. The Spanish surrealist artist Salvador Dali referred to drawing as the honesty of art, saying "there is no possibility of cheating. It is either good or bad." With that in mind, I hope you enjoy my renditions of the Messier objects and find them useful in lending perspective to your observations.

THE FINDER CHARTS

The constellation map at the back of the book is sufficient for locating most, if not all, of the Messier objects, which are fairly bright and easy to spot once you're looking in the general vicinity. But the smaller-scale finder charts provide additional detail. They can be particularly helpful in identifying other objects that lie near to the Messier objects. The scale and magnitude limit of the finder charts vary. For some, a smaller scale (and higher magnitude limit) may have been needed to distinguish several close-together objects, whereas a wider scale (and lower magnitude limit) was appropriate to show objects separated by greater distances or to show an object's location relative to a large pattern of stars in a constellation.

The following symbols are used to represent the different object types on the finder charts:

▢	Diffuse nebula	⬡	Open cluster	⬭	Galaxy
▪	Dark nebula	⊕	Globular cluster	●	Variable star
⟡	Planetary nebula	△	Supernova Remnant	●–	Double star

MESSIER'S OBJECT DESCRIPTIONS

Messier's own descriptions of the objects he catalogued are included in chapter 4. The translation of the French catalogue in the *Connaissance des Temps* for 1874 was expertly done by Storm Dunlop, author and translator of numerous books on astronomy and a fellow of the Royal Astronomical Society. Storm gained access to the catalogue in the library of the RAS in

London, where it is preserved. His is the most precisely interpreted and smoothest-reading English translation of the catalogue that I have seen, and I am grateful to him for bringing his linguistic gifts and knowledge of astronomy to bear on this important task.

There are a few things to note about Messier's catalogue and its translation. His object descriptions appeared on right-hand pages, while the facing left-hand pages contained columns listing the object number, the right ascension and declination, his estimate of the object's angular diameter, and the date of his observation (which I have put in brackets preceding each description). This edition of the catalogue, the last one published before Messier's death, included 103 objects; numbers 104 through 110 were added later.

Messier used the third-person grammatical form when referring to himself, saying of M32, for example, "M. Messier saw it for the first time in 1757, and has not noted any change in its appearance." This style was not imposed by the translator. Bracketed [ ] words or phrases within the descriptions were added in translation for purposes of clarification. Messier also made frequent use of the term "ordinary telescope," which has been translated literally by others in the past. The more correct meaning, "simple refractor," is used here. The implication is that the telescope used was not a compound (achromatic) refractor. When Messier describes an observation made with a "one-foot telescope," or a "simple three-foot telescope," he is referring to the scope's *length*, not its aperture. Messier's use of the word "parallel" was translated literally, e.g., "the cluster is close to Antares and on the same parallel." Dunlop suggests that Messier might have meant something more like "zone of declination," because he often referred to an object being on the same parallel, but slightly above (or below) it. Messier occasionally refers to an object being plotted on the English *Atlas Céleste*, by which he means an original English edition of Flamsteed's *Atlas Coelestis*. A French-language edition, *Atlas Céleste*, was published in 1776, several years before Messier's catalogue.

The Latin names of constellations are given throughout, rather than the vernacular names that Messier used. Punctuation has been modernized.

4 The Messier objects

M1

Crab Nebula
NGC 1952
Type: Supernova Remnant
Con: Taurus
RA: 5^h 34^m.5
Dec: + 22° 01'
Mag: 8.4; 8.0 (O'Meara)
Dim: 6' × 4'
Dist: ~6,500 l.y.
Disc: John Bevis, 1731

MESSIER: [Observed 12 September 1758] Nebula above the southern horn of Taurus, which does not contain any stars. Its light is whitish and elongated like a candle flame. Discovered when observing the comet of 1758. See the chart of this comet, *Mémoires de l'Académie 1759*, page 188; observed by Dr. Bevis around 1731. It is plotted on the English *Atlas Céleste*.

NGC: Very bright and large, extended along position angle approximately 135°; very gradually brightening a little toward the middle, mottled.

Messier had no idea his first catalogued object would be among the most intriguing in the heavens. M1, the Crab Nebula, is one of 100 or more known supernova remnants in our galaxy – a corpse of a star that experienced a fast life and a violent death. A supernova explosion is the final stage in the life of a star some 15 times more massive than our sun. Such a *red supergiant* star (like Betelgeuse, in the shoulder of Orion) voraciously consumes its nuclear fuel in about 10 million years (100 times faster than the sun). When the star's thermonuclear energy is exhausted, its earth-size core collapses under the force of gravity and, within seconds, shrinks until the core's density equals that of an atomic nucleus. Unable to contract further, infalling gas rebounds off the resistant core. A quarter-second later, the star ends its life in a fantastic explosion, the peak energy of which can rival that of its host galaxy.

The Crab is the remains of a cataclysmic stellar explosion that occurred in our own Milky Way galaxy in A.D. 1054. So powerful and so close (approximately 6,500 light years) was the blast that Chinese sky-watchers described it as a "guest star" in the annals of the Sung dynasty. It shined as bright as Venus in the daytime sky, appeared reddish white, and was observed for 23 days.

As the only supernova remnant in Messier's catalogue, M1 warrants special attention. In small telescopes it is a 6' × 4' irregular patch of nebulosity situated a little more than 1° northwest of the 3rd-magnitude star Zeta (ζ) Tauri, a hot, blue subgiant star. The nebula is surprisingly easy to

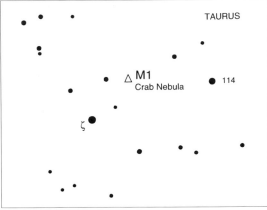

see with 7×35 binoculars (amazing, if you consider that nearly two millennia have passed since the explosion). Curiously, most catalogues fail to provide a precise visual-magnitude estimate for this very famous object. In the *Messier Album*, Mallas and Kreimer give it a magnitude of "8 or 9"; Burnham says "about 9." Jones, in his work *Messier's Nebulae and Star Clusters*, gives it a more precise magnitude, 8.4. All these estimates seem a trifle too faint.

At 23× the Crab Nebula shares the field with Zeta Tauri, and the nebulous glow looks much like the ghost image of that star. But the nebula, which measures roughly 11 light years by 7½ light years, is so much more enormous. It is composed of three parts: a 16th-magnitude pulsar (a rotating neutron star), an inner bubble of material (a powerful wind of radiating particles bound to the object's magnetic field), and an outer shell of dense material released in the supernova explosion.

With a glance at low power, the nebula's midsection appears pinched. A longer look will reveal two halves slightly askew or misaligned, as if two plates along a tectonic fault had suddenly slipped. At higher magnifications the nebula looks patchy with three distinct sections running southeast to northwest. The southern and middle sections are of similar brightness, while the northern one is smaller and much fainter. When photographed in polarized light, the nebula reveals a similar trilobate or serrated aspect, indicating the existence of very strong magnetic fields. Has anyone with a large telescope ever tried to view M1 with polarized glass, like the rotating polarizer used in terrestrial photography?

The Crab's eastern edge contains a prominent notch, or bay, accentuated by a long filament flowing to the southeast. This filament meanders through the nebula's midsection to the western side, where it extends beyond the main body. Look carefully and see if you can detect a gray river adjoining the filament, which visually separates the southern and middle portions. An enhancement at the northern boundary of the southern section abuts the gray river. It looks like an elongated patch of nebulosity (running east to west) with a possible dual nature.

When I first saw this patch, I wondered if this might be the blended image of the Crab's rapidly spinning neutron star and its equally bright neighbor star. Alas, at magnitude 16, the neutron star (whose pulses of x-ray, optical, and radio energy every 0.033 second illuminate the nebulosity) is too faint for such a small instrument. The enhancement, however, is just south of the Crab's true double heart. Interestingly, this patchy feature also shows well in polarized-light images.

I did view the neutron star through a 20-inch Dobsonian at the 1990 Winter Star Party in the Florida Keys. From the best sites, the pulsar can be seen in a 10-inch telescope with high-quality, unobstructed glass. That night in Florida I also noticed that the main nebulous body is surrounded by a fainter glow composed of a network of fine filaments. Rosse first noted the Crab's filamentary structure in 1844. And though it is commonly stated that large telescopes are needed to bring out these delicate features, they can be glimpsed in a 4-inch glass. The problem is not one of aperture but of sky background and contrast.

Data from Hubble Space Telescope observations of the Crab Nebula in 1994 reveal that these filiments are cloaked in a glowing plasma. In a paper published in the 1 January 1996 issue of the *Astrophysical Journal*, J. Jeff Hester (Arizona State Unversity) and his team suggested that the glowing filaments develop where fast-moving plasma from the Crab's inner bubble of material pushes on the dense outer shell. HST data also show regions of magnetic instability, where fingers of plasma pour back into the inner bubble as it pushes outward. Some astronomers have suggested that there is also an invisible element to M1 – more material beyond the Crab's visible extent – through which this inner bubble is expanding. The bubble, then, might be sweeping up this unseen material and channeling it into the visible, fingerlike structures resolved by HST. If so, this finger formation in the filiments is an ongoing process.

Finally, have you ever wondered how M1 derived its nickname, "the Crab"? Jones explained that the name stems from a drawing made in 1844 based on observations with the 36-inch Rosse reflector at Birr Castle in Ireland. The drawing bears some resemblance to a horseshoe crab, though, as Jones pointed out, it really looks more like a pineapple. Rosse virtually repudiated that sketch, yet the moniker remains. Interestingly, in the 4-inch the nebula's main body looks very much like a crab or lobster claw, so I derive some pleasure in having found a way to preserve this intriguing historical interpretation.

M2

NGC 7089
Type: Globular Cluster
Con: Aquarius
RA: 21^h 33^m.5
Dec: –0° 49′
Mag: 6.6; 6.3 (O'Meara)
Dia: 16′
Dist: 37,000 l.y.
Disc: Jean-Dominique
Maraldi II, 1746

MESSIER: [Observed 11 September 1760] Nebula without a star in the head of Aquarius. The center is bright, surrounded by circular luminosity; it resembles the beautiful nebula that lies between the bow and the head of Sagittarius. It is clearly visible with a two-foot telescope, set on the same parallel as α Aquarii. M. Messier plotted it on the chart showing the path of the comet observed in 1759, *Mémoires de l'Académie 1760*, page 464. M. Maraldi saw this nebula in 1746, while observing the comet that appeared in that year.

In autumn, the vast summer Milky Way slips drowsily into the western horizon after sundown. Hours will pass before mighty Orion and other bright winter constellations rise in the east. Looking overhead, we now peer straight out of the galaxy – away from the crowded stellar cities of the Milky Way's arms – into a suburb of stars whose residents include some of the most inconspicuous constellations in the night sky; chief among them Aquarius, the Water Bearer. It is largely indefinable, and its faint stars must compete with light pollution. Nevertheless, Aquarius contains a secret treasure well worth hunting for – the spectacular globular cluster M2.

Like M1, M2 was a chance discovery for Messier. He happened upon it while looking for a comet in 1760, but Jean-Dominique Maraldi II in Paris had noticed it 14 years earlier. Maraldi suspected this nebulous object was a cluster, but his reasoning was clearly hypothetical: since he could not detect any field stars surrounding this unresolvable haze, the haze itself, he deduced, must be made of innumerable stars too faint to be seen.

The popular nineteenth-century British observer Adm. William Henry Smyth seemed especially fond of M2: "This magnificent ball of stars condenses to the centre and presents so fine a spherical form that imagination cannot but picture the inconceivable brilliance of their visible heavens to its animated myriads." Although Smyth's words are a little hard to follow, it's clear that he was thoroughly impressed.

To find M2 without setting circles or automatic devices, you need a proper knowledge of the constellations, because the cluster lies in a region relatively barren of stars. It's easiest to first locate 2.4-magnitude Epsilon (ε) Pegasi (the nose of the mythical winged horse, Pegasus) and 3.9-magnitude Alpha (α) Equulei, nearly a fist width to the southwest. From the midpoint between Alpha Equulei and Epsilon Pegasi, look about 5° to the southeast and you'll find three 6th-magnitude stars – 25, 26, and 27 Aquarii. Just south of 25 and 26 Aquarii, a string of binocular stars become increasingly bright to the southwest. Follow them because they point directly to M2. This 6th-magnitude cluster stands out prominently as a round, hazy patch in binoculars. It can also be seen, without too much difficulty, with the unaided eye, although you may have to use averted vision.

Except for a 10th-magnitude star about 5′ to its northeast, M2 appears rather lonely at low power. The globular lies far from the plane of the Milky Way, so its telescopic surroundings appear relatively devoid of bright background stars. But the visual impact of the cluster itself makes up for that deficiency. At 23× this 170-light-year-wide swarm of 100,000 suns – replete with yellow and red-giant stars about 13 billion years old – displays a very tight, starlike center surrounded by a yellow outer core that has a diffuse, pale-blue halo. The 7-mm eyepiece resolves a sprinkling of stars, the brightest of which are about 13th magnitude. But add a Barlow lens and, with averted vision, the dotted haze becomes a multitude of

NGC: Very remarkable globular cluster, bright, very large, gradually pretty much brighter toward the middle, well resolved into extremely faint stars.

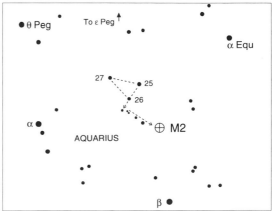

faintly sparkling gems. The strong illusion of a central star at low power vanishes with moderate to high magnifications, when the globular is transformed into a mysterious ball of starlight and shadows.

If you can let your gaze alight a moment on the individual shadows, the illusion of M2's spherical nature will be shattered. The outer envelope takes on a peculiar north–south asymmetry with an explosion of starlight at its fringes. Now the mystery shadows seem to blow out from the center with the star streams, forming spidery arms. One particularly obvious, rogue shadow has been noted by several observers. It slices through the northeast section of the outer halo and runs northwest to southeast. The neighboring 10th-magnitude star is a perfect guide, because the shadow lies roughly halfway between it and the cluster's center.

Look at M2 several times over the course of two weeks, because it contains a prominent variable star, which can greatly impact the cluster's appearance. A French amateur astronomer named A. Chèvremont discovered this pulsating star in 1897. Located on the eastern edge of the cluster, just slightly north of center, Chèvremont's variable ranges in brightness from magnitude 12.5 to 14.0 over about 11 days (that period may fluctuate). At maximum, Chèvremont's variable should be within range of a good 2-inch telescope.

NGC 5272
Type: Globular Cluster
Con: Canes Venatici
RA: 13ʰ 42ᵐ.2
Dec: +28° 22'
Mag: 6.3; 5.9 (O'Meara)
Dia: 19'
Dist: 32,000 l.y.
Disc: Messier, 1764

MESSIER: [Observed 3 May 1764] Nebula discovered between Boötes and one of Hevelius's Hunting Dogs [Canes Venatici]. It does not contain any stars, the center is bright, and its light decreases imperceptibly [away from the center]; it is circular. Under a good sky it can be seen with a one-foot telescope. It is plotted on the chart of the comet observed in 1779, *Mémoires de l'Académie* for that year. Observed again 29 March 1781, still as fine.

NGC: A very remarkable globular; extremely bright and very large; toward the middle it brightens suddenly; it contains stars which are 11th magnitude and fainter.

M3 is another challenging object for beginners. Not only is it a great naked-eye challenge, but, as discussed in chapter 3, the stars in the cluster itself can be used to test telescopic vision (see the chart on page 33). Although it is located in Canes Venatici, it's best to use the stars of Boötes and Coma Berenices as guides. M3 is just northeast of an orange 6th-magnitude star about two finger-widths due east of Beta (β) Comae Berenices and 10° (one fist) north and slightly west of Eta (η) Boötis. Once located in binoculars, this cluster is easier to see with the naked eye than M2 for several reasons: it is a half-magnitude brighter, there is a 6th-magnitude guide star next to it, and it resides much higher in the northern hemisphere sky. Here's the catch: M3 lies so close to the guide star it may be hard to resolve the two! Both Skiff and I achieved this from the dark skies of Texas, and I have no difficulty seeing M3 with the unaided eye from Hawaii.

Although M3 lies in a relatively star-poor field, it is contained in an acute triangle of 9th-magnitude stars, and an obvious topaz star borders the cluster to its northwest. With a glance using 23×, the 19'-wide glow displays a stellarlike core surrounded by a gradually fading halo. But look

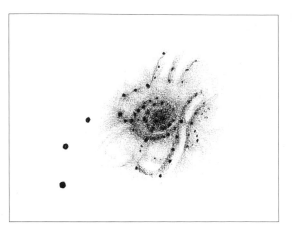

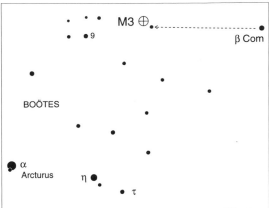

closely at the bright central point, because it appears bounded by a very tight inner shell with a distinct *peach* color. (I have noticed that when the cluster is low in the sky, this inner region looks yellow.)

The observed color of globulars is open for debate. Some astronomers argue that the colors are merely optical illusions. But I find that at very low magnifications most bright globulars show distinct, albeit *faint,* colors – mostly shades of yellow and blue. Other observers have independently seen these tints, and true-color images of globulars reveal them as well. Admittedly, green tints, which some globulars appear to have, may be illusory, being a combination of the yellow and blue casts. Not surprisingly, I have also found that the blue and green colors of globulars seem to vanish with notable increases in magnification.

Continuing at low power, M3's peach-tinted inner shell is framed by a larger and fainter mantle, which shines weakly with an aqua luster (more greenish with some blue). Beyond this is an even larger envelope, whose overall color is blue-green. M3 should certainly be rated as one of the most colorful globulars for small telescopes in the Northern Hemisphere. At the least, its color scheme is intriguing and apparent.

At 72× the cluster metamorphoses into a finely resolved sphere with an uncanny three-dimensional quality; I feel like I'm looking down on a snowball melting on black ice. The cluster's core appears bulbous and its edges flat and sprawling. Moderate power also reveals a tantalizing string of stars that connects the cluster to the topaz star about 10′ to the northwest. When seen in a simple inverted telescope, M3 seems to dangle from the star like an earring.

Some observers, such as the late Walter Scott Houston – whose Deep Sky Wonders column in *Sky & Telescope* magazine gave my generation of observers constant food for thought – have noticed that the central region is skewed to the west. I too see this, but believe it to be an illusion caused by a curious gathering of stellar clumps west of the cluster's true center. A series of westward-extending arms adds to this illusion. Otherwise, medium and high powers resolve the cluster into a myriad of ultrafine stars, like tiny emerald chips scattered on a golden carpet. So uniform and gauzelike is the light from each individual star enveloping the bright core that I imagine it as a candle burning behind a curtain of green lace.

A good project for someone with a large telescope and a photometer might be to periodically check the brightness of stars in M3. Of the half-million suns shining in this venerable cluster, more than 180 are known to be variable. (For comparison, M2 has about 20 known variable stars.)

Visual observers should look for the mysterious dark spots that inhabit M3's nuclear region. Lord Rosse first noted them from his observatory at Birr Castle in Ireland as "small, dark holes"; they show up well on high-resolution photographs.

M4

Cat's Eye
NGC 6121
Type: Globular Cluster
Con: Scorpius
RA: 16ʰ 23ᵐ.6
Dec: –26° 31′
Mag: 5.4
Dia: 35′
Dist: 6,800 l.y.
Disc: Philippe Loys de Chéseaux, 1746

Riding high in the springtime sky, globular cluster M4 awaits the inquisitive gaze of any amateur curious about the limits of vision. As you probe the depths of this cluster, keep in mind that you are looking at an object 70 light years in diameter and roughly 10 billion years old, a senior member of our Milky Way Galaxy.

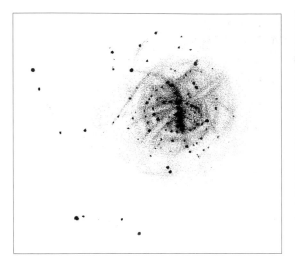

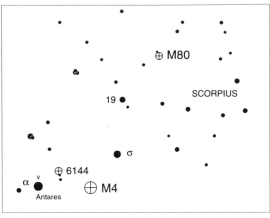

MESSIER: [Observed 8 May 1764] Cluster of very faint stars. With a small telescope it looks like a nebula. This star cluster is close to Antares and on the same parallel. Observed by M. de la Caille, and included in his catalogue. Observed again 30 January and 22 March 1781.

NGC: Cluster, with 8 or 10 bright stars in a line . . . , readily resolved.

Although brighter than either M2 or M3, M4 lies so close to the brilliant red supergiant Antares (1½° to the west) in Scorpius that glare from the 1st-magnitude beacon all but ruins a clear and consistent naked-eye view of it. With binoculars, the cluster appears very bright and round. Telescopically, at low power the 35′ disk of M4 immediately resolves into a loose throng of 11th-magnitude stars, which betrays the cluster's overall roundness and symmetry; bisecting this sphere is a spindle of about a dozen 10th- to 12th-magnitude stars running north to south. With medium power this ridge appears more obvious and elongated. Here is a cat's keen eye, with its slit of a pupil, peering at you through the dark vapors of night. Although the Cat's Eye moniker is my own (which should not be confused with NGC 6543, the Cat's Eye Nebula, an 8th-magnitude planetary), I was by no means the first to see the "pupil" feature. William Herschel, discoverer of Uranus, described M4 in 1783 as a "ridge of 8 or 10 pretty bright stars," and Smyth saw it "running up to a blaze in the centre."

Behind that stunning slit of starlight is a swollen haze of unresolved cluster members. Just when I feel I can resolve a faint member, that moment passes and I catch sight of another "spark," until my eye spies another, and so on. One can spend hours being hypnotized by the comings and goings of these fleeting flashes. Burnham had a similar impression of M4 after he observed the cluster through the Yerkes 40-inch refractor. He writes:

> It happened that a few days before, I had obtained a small but fascinating device called a spinthariscope in which the effect of radioactivity is made visible to the eye; ever since I have mentally associated M4 with alpha particles. . . . The observer, after a period

of dark adaptation, looks into the lens [of the spinthariscope] to see a view resembling M4 "brought to life" with hundreds of microscopic "stars" blazing up and vanishing every second.

Shock waves of dark lanes radiate from the needlelike hub. Defocus the telescope slightly to follow these dark ripples. So many populate the region that with some imagination you can envision a shattered glass photographic plate of M4, or a cobweb silhouetted against an approaching swarm of lightning bugs, or, better yet, diamonds snatched up in a loose black net, with many leaking through the fibers into limitless space.

The outlying northern section appears to sink gradually into a black ocean, and if you look hard enough, you can perceive a submerged shoal of faint stars, a celestial reef. Many of the stars outside the core form concentric horseshoe-shaped patterns that loop northward from the south. Stellar arms fly off in various directions. In this way, the cluster looks more like a collision of two pinwheel galaxies such as M33 or M100, with flat, broad, and loose spiral arms.

While in the area, check out the tiny 9th-magnitude globular cluster NGC 6144 just $\frac{1}{2}°$ northwest of Antares. Its feeble glow reminds me of a ghost image of Sigma Scorpii, a hot blue-giant (B1) star that shines at 3rd magnitude. NGC 6144 is a highly neglected object, hidden in the glare of Antares and overshadowed by M4.

M5

NGC 5904
Type: Globular Cluster
Con: Serpens (Caput)
RA: 15^h 18^m.5
Dec: +2° 04'
Mag: 5.7
Dia: 22'
Dist: 25,000 l.y.
Disc: Gottfried Kirch, 1702

MESSIER: [Observed 23 May 1764] Beautiful nebula discovered between Libra and Serpens, close to the sixth-magnitude star Flamsteed 5 Serpentis. It does not contain any stars; it is round, and it may be seen very well under a good sky with a simple one-foot refractor. M. Messier plotted it on the chart for the comet of 1753, *Mémoires de l'Académie 1774*, page 40. Observed again 5 September 1780, and 10 January and 22 March 1781.

NGC: Very remarkable globular cluster, very bright, large, extremely compressed in the middle, stars from 11th to 15th magnitude.

The fifth object in Messier's catalogue is a powerful and dynamic sight in small telescopes. Even at low power it is a slightly stellar conflagration with a blazing heart. A wide and loose, slightly elliptical exterior becomes increasingly tight toward a starlike center. The cluster looks as if it is collapsing under the force of gravity, triggering atomic reactions in its core. And with a 7-mm, the entire cluster seems electric, bursting with fiery sparks.

Now, contrast that with Mary Proctor's musings on the same globular, which she viewed through the world's largest refractor, the 40-inch Clark at Yerkes Observatory. The description is from her 1924 book, *Evenings With the Stars:* "Myriads of glistening points shimmering over a soft background of starry mist, illumined as though by moonlight. . . . For a few blissful moments, during which the watcher gazed on this scene, it

Deep-Sky Companions: The Messier Objects

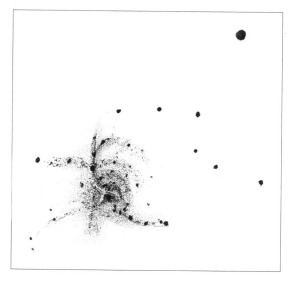

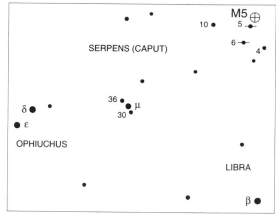

suggested a veritable glimpse of the heavens beyond." Observing is indeed a highly personal experience; sometimes we see what we feel.

To find this easy naked-eye object, look 20′ northwest of the fine double star 5 Serpentis (a golden 5th-magnitude star with a 10th-magnitude ashen companion a little more than 2′ away). Color – namely, a straw interior with a powder blue exterior – is immediately obvious in M5. Few accounts I've seen mention the obvious curved wings of 12th-magnitude stars stretching northeast to south from the core. This region reminds me of an airborne owl, whose feathered wings shimmer with reflected light. At medium power, the wings are amazingly well defined. Luginbuhl and Skiff had a different impression, one of a "rich open cluster superposed on a bright galaxy."

At 130×, the nuclear region appears brightest to the north, with a faint stubby wing just to its west. Toward the southeast, stars spiral or fan out in long arms – a pattern Rosse noticed in 1875. These features seem to originate from stellar kinks along a central bar, the kinks themselves being chance associations of stars.

As you gaze at this incredible, 13-billion-year-old object – without question the finest globular cluster in the northern sky for small telescopes – try contemplating the following. M5 is superposed on the edge of a faint, distant cloud of galaxies, which is composed of several groups of galaxies – a mind-boggling aggregate with some 200 members per square degree. Four degrees (about two finger-widths) due west of M5, at least eight galaxies surround the 4th-magnitude star 110 Virginis (not shown on the finder chart). All are visible at low power, but they are best seen at moderate magnifications. Take your time studying this region and confirm the galaxies on your star atlas.

M6

Butterfly Cluster
NGC 6405
Type: Open Cluster
Con: Scorpius
RA: 17^h 40^m.3
Dec: −32° 15′.5
Mag: 4.2
Dia: 20′
Dist: 1,585 l.y.
Disc: Philippe Loys de Chéseaux, 1746; Giovanni Batista Hodierna saw it before 1654, and Claudius Ptolemy recorded it in the second century A.D.

MESSIER: [Observed 23 May 1764] Cluster of faint stars between the bow of Sagittarius and the tail of Scorpius. To the naked eye this cluster appears to form a starless nebula, but even the smallest instrument shows it to be a cluster of faint stars.

NGC: Cluster, large, irregularly round, loosely compressed, stars from 7th to 10th magnitude.

Located about 4° northeast of the blue subgiant star Lambda (λ) Scorpii, the easternmost of the Scorpion's two stinger stars, are M6 and M7, two of the most striking details in the most dramatic naked-eye expanse in the heavens – the hub of our galaxy. These clusters are among the more brilliant Messier objects in the night sky. (M7, the southernmost Messier object, is arguably the brightest spot in the entire visual Milky Way.) They all but dominate the summer Milky Way and erupt like distant fireworks. From dark southerly locations, both M6 and M7 are visible to the naked eye as puffs of smoke even under full moonlight!

Personally, I cannot look upon one of these clusters without immediately trying to contrast it with the other. And though they are separated by 3½° of sky, I view them as a double cluster, fraternal twins. Actually, the two clusters are some 805 light years apart, but aren't we entitled to enjoy them as we please?

The fainter of the two, M6, is famous for its butterfly pattern, which looks more like a 1960s butterfly motif (the kind worn on bell-bottoms) than, say, a swallowtail, but nevertheless, the imagery is quite striking. It's best seen at low magnification, when the butterfly's antennae (see the drawing) add a tad more detail to the design. Aside from the butterfly, look to the northwest where a definite V-shaped pattern of ice blue stars dominates the telescopic view; this grouping shows up especially well with moderate powers.

M6 contains 126 stars to magnitude 14, about half of which are magnitude 11 or brighter; the entire cluster contains more than 330 suns. Most of the stars are dazzlingly crisp blue-white gems. The major exception is orange BM Scorpii, a semiregular variable whose light fluctuates from magnitude 5½ to 7 in about 850 days. See if you can detect BM with the unaided eye; it marks the northeast tip of the butterfly's wing. There should be no problem doing this when it shines at maximum. If you

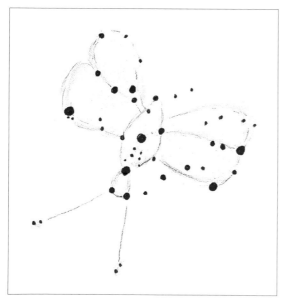

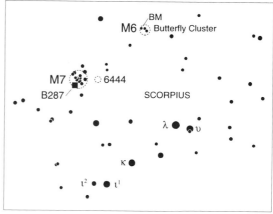

M7

NGC 6475
Type: Open Cluster
Con: Scorpius
RA: 17^h 53^m.8
Dec: −34° 47′.1
Mag: 3.3; 2.8 (O'Meara)
Dia: 80′
Dist: 780 l.y.
Disc: Claudius Ptolemy,
second century A.D.

MESSIER: [Observed 23 May 1764] A larger cluster of stars than the previous one [M6]. This cluster appears to be a nebula to the naked eye; it is not far from the previous one, lying between the bow of Sagittarius and the tail of Scorpius.

NGC: Cluster, very bright, pretty rich, loosely compressed, stars from 7th to 12th magnitude.

succeed, you will have resolved a star in a cluster 1,585 light years distant! Actually, the cluster contains four widely separated 7th-magnitude stars, which under ideal conditions could be detected with the unaided eye.

Use your binoculars to study the Milky Way region surrounding M6. Then train them on nearby M7 and its surroundings. See anything noticeably different? M6 is surrounded by as much darkness as M7 is by

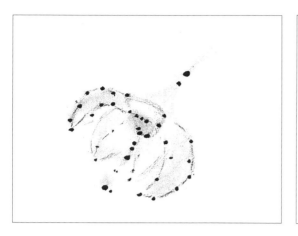

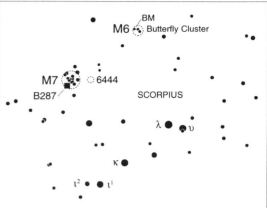

light! This aspect is even more apparent in the 4-inch at 23×. In fact, when sweeping from M7 to M6, I noticed that M7 lies in the fattest part of a bright wedge of Milky Way – a rich star cloud – that points to M6. Tiny M6 sits like an island off the glittering sand of this Milky Way beach. With your binoculars or your telescope at low power, look at the region immediately surrounding M6; you should see a wall of open clusters running from the northeast to the southwest, including NGC 6451, 6425, 6416, and 6404. A little south of west is another loose sprinkling of stars, NGC 6383.

I am not sure why over the last few centuries more observers didn't sing praises about M7. In the seventeenth century, Giovanni Hodiera wrote merely, "Counted 30 stars." An apparently unimpressed Lacaille described M7 from the Cape of Good Hope as a "group of 15 or 20 stars very close together in a square." And in his 1864 catalogue, John Herschel characterized it as "a cluster, very bright, pretty rich, little compressed."

Perhaps most of the observations were made from temperate latitudes in the Northern Hemisphere. From the northern tropics and the Southern Hemisphere, there are few grander naked-eye sights than this magnificent stellar wonder. M7 blazes against the river of the Milky Way like the head of some great lost comet.

M6 is estimated to be 100 million years old; M7 is more than twice that age. Compared to the icy-hued stars of M6, those in M7 appear more golden, like sun-drenched hay. M7 has about 80 stars, all of which shine brighter than 10th magnitude, and a dozen of these are brighter than 7th magnitude. The challenge, then, is to resolve this cluster with the naked eye. Clearly half of these stars are 6th magnitude and should be discernible from suburban skies. If you live under a very dark sky, try during the quarter moon, which will diminish the cluster's background noise of fainter, unresolved stars. Another trick is to catch M7 in the morning or evening twilight when its light just starts to fade or emerge, respectively.

With binoculars M7 looks like a gem-studded cross inside a double halo of similarly bright stars. With imagination, it is possible to perceive the cross and the two haloes all appearing on different planes. Defocus the binoculars slightly and see if you can detect the black slash running along the western side of the cross's long axis. That dark wound in the Milky Way points to Barnard 287, a pond of dark matter due south of the cluster.

Telescopically, M7 resembles a cosmic flower opening in the morning mist of the Milky Way, the long axis of the cross being the flower's stamen, and the haloes its petals. NGC 6444, a modest open cluster to the west, is a burst of pollen. For a challenge, try to pick out the tiny, 10th-magnitude globular cluster NGC 6453 1° northwest of M7's center.

M8

Lagoon Nebula
NGC 6523
Type: Diffuse Nebula and
Cluster
Con: Sagittarius
RA: $18^h 03^m.8$
Dec: −24° 23′
Mag: 4.6; 3.0 (O'Meara)
Dim: 90′ × 40′ (nebula)
Dia: 14′ (cluster)
Dist: 5,200 l.y.
Disc: John Flamsteed, ~1680

MESSIER: [Observed 23 May 1764] A cluster of stars that appears to be a nebula when observed with a simple three-foot refractor; with an excellent instrument, however, one sees only a large number of faint stars. Near this cluster there is a fairly bright star, which is surrounded by a very faint glow; this is the seventh-magnitude star Flamsteed 9 Sagittarii. The cluster appears to be elongated in shape, extending from northeast to southwest, between the bow of Sagittarius and the right foot of Ophiuchus.

NGC: A magnificent object, very bright, extremely large and irregular in shape, with a large cluster.

Look about 5° (two finger-widths) west and slightly north of 3rd-magnitude Lambda (λ) Sagittarii, the orange *K*2 III star that marks the top of the famous teapot asterism in Sagittarius. Very conspicuous to the naked eye, M8 appears as a large curdle of galactic vapor off the western edge of the Milky Way that rises from the teapot's spout. The emission nebula, powered by the radiative energy of the very hot 6th-magnitude star 9

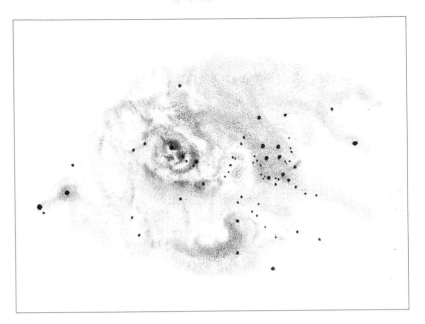

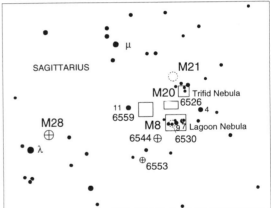

Sagittarii, 9th-magnitude Herschel 36, and possibly some obscured stars, is complemented on its eastern side by NGC 6530 – a loose spritz of 113 very young suns, all of which are probably intimately associated with the M8 nebulosity that enshrouds them in loops and swirls. I was surprised at my magnitude estimate of 3.0 for M8, which was derived by defocusing my eyes and comparing the nebula's light to that of Lambda and Gamma (γ) Sagittarii. Other sources rate M8's magnitude anywhere from 4.6 to 6.0. The discrepancy could arise from the object's low altitude and from the dimming effects of light pollution. But Brent Archinal of the U. S. Naval Observatory believes these fainter estimates are more accurate for the cluster itself and not the nebulosity.

Here is a simple visual check. Look at M8 and then compare it to M6, which shines at magnitude 4.4. It's obvious: M8 is much brighter. Another check is to see which emerges first from the evening twilight.

The word "Lagoon" was probably first used in association with M8 by Agnes M. Clerke in her 1890 work entitled *The System of the Stars*. The name refers to the striking east-to-west–running black furrow that divides the nebula's brightest regions, which looks more like a channel than a lagoon. The channel is but a part of the lagoon, which is really a V-shaped body of darkness that embraces a bright island of nebulosity to the west and a curved "sandbar" of nebulosity to the east. If you look long and hard enough, you should actually see two lagoons: a "deep" and narrow inner one, and a wider, "shallower," outer one. Can you see the skeletal-like fingers of nebulosity to the south of the Lagoon's main dark channel? Anyone familiar with the classic horror film *Creature From the Black Lagoon* might find some similarity between these nebulous fingers and those webbed hands of the Creature.

The dark channel is so prominent at low power that it must be the most dramatic example of dark nebulosity in any deep-sky object visible in small telescopes. The entire nebula appears to be caught between worlds of darkness and light. John Herschel described it as a "collection of nebulous folds and matter surrounding and including a number of dark, oval vacancies." Use moderate power and averted vision to see the dark nebulosity that cages the bright nebula.

Particularly delicate at $23\times$, the nebulosity and its myriad dark lanes look like a frozen flower petal that has fallen to the ground and shattered. If you mentally erase the nebulosity, you should see a crossbow of seven prominent stars – the skeleton of this Messier object. To the naked eye and in binoculars, the nebula and the staff of the crossbow constitute what I immediately recognize as M8, and this is what Messier saw. The NGC 6530 designation applies to the stars in the eastern part of the bow. Concentrate on the position of NGC 6530 in the nebula and see if it doesn't seem to weigh down the southern part of the nebulosity. To me, it is a very incongruous sight. I remedy that illusion by creating another: I like to imagine that the stars of the crossbow and the cluster are foreground objects and that the nebulous matter is far behind them.

Reserve plenty of time to study the heart of M8, just $3'$ west-south-west of 9 Sagittarii. It is easily recognizable as a very dense region of nebulosity surrounding the star Herschel 36. With moderate magnification, the region shines with a dusky yellow or straw color. Its main features are two bright knots (southeast and northeast of Herschel 36) and a "wishbone" of dark lanes to the northwest. The bright knots, whose tapered ends join in the middle, make up the famed "hourglass" nebula – a mysterious source of radio emission. But NASA's Hubble Space Telescope is rapidly solving many deep-sky mysteries. In our own galactic neighborhood, it is opening doors to secret cosmic gardens (like the "hourglass" region in M8) and revealing the wonders within. HST's portrayal of the "hourglass" is nothing short of spectacular. In a region only one-half light-year long, interstellar "twisters" – eerie dark funnels – project from nascent clouds like tornadoes on Earth. These twisters are silhouetted against the brilliant backdrop of ionized gas in the "hourglass." The large difference in temperature between the hot surface and cold interior of the "hourglass" clouds, combined with the pressure of starlight, may produce strong horizontal shear to twist the clouds into their tornado-like appearance. A curious knot of nebulosity appears isolated from the "hourglass" to the north of Herschel 36, could it be part of the "hourglass" separated by these dark twirling clouds?

Just to the north of this region you will encounter a long, diffuse swath of nebulosity that arcs over all these features. To the west, a faint nebulosity veils the 5th-magnitude $F5$ star 7 Sagittarii and its fainter neighbor further to the west; a dim bridge of cloud connects them to M8.

Photographs, such as the one at the opening for this object, reveal several tiny, round globules of dark matter known as Bok globules after the late Harvard University astronomer Bart Bok, who interpreted the mysterious dark nebulae, which were first recorded in photographs by turn-of-the-century astronomer E. E. Barnard. These dense, obscuring clouds have diameters of about one-third of a light year and larger and are believed to be sites of star formation. Bok globules are similar to, but different from, the dense star-spawning nebulae called "evaporating gaseous globules" (EGGs) that the Hubble Space Telescope imaged in M16, a nebula and cluster in Serpens. Bok globules are about 50 times larger than EGGs, whose placental clouds of dust and gas are about the size of our solar system. Like the EGGs, Bok globules undergo photoevaporation, the process in which intense ultraviolet radiation from young, hot O- and B-type stars blows away surrounding dust and gas, uncovering sites of star formation.

Bok globules and EGGs are too faint to be seen in small telescopes, but I did unknowingly record a *mock* Bok globule. It is actually Barnard 88, a cometlike dark nebula in the northeast section of the outer cloud (see the photograph). Visually, it appeared as a notch or bay in that swirling swath of brightness. But when I compared the drawing to the photograph, I saw that the location of the notch matched that of Barnard's black "comet."

M9

NGC 6333
Type: Globular Cluster
Con: Ophiuchus
RA: $17^h 19^m.2$
Dec: $-18° 31'$
Mag: 7.8
Dia: 11'
Dist: 22,500 l.y.
Disc: Messier, 1764

MESSIER: [Observed 28 May 1764] Nebula without a star, in the right foot of Ophiuchus; it is circular and its light faint. Observed again 22 March 1781.

NGC: Globular cluster, bright, round, extremely compressed middle, well resolved, stars of 14th magnitude.

As you hunt down the Messier objects, try to forget any preconceived notions you might have about their appearance. For example, it is easy to assume that all Messier globular clusters will at first appear as tight, fuzzy balls of starlight. But M9 is clearly an exception. It immediately looks different, as though someone has tried to erase it from the sky.

M9 glows about 3½° southeast of 2.5-magnitude Eta (η) Ophiuchi, which is nearly 15° (a fist- and two finger-widths) to the northeast of brilliant Antares in Scorpius. With an angular diameter of 11', M9 appears as a tiny patch of light in binoculars. At 23× it remains dwarfish and looks highly compressed. Its appearance is all the more paltry if you happen to take in the more dazzling Ophiuchus globulars M10 and M12 first.

Curious as to why M9 looks unusual, I swept the region at 23× and was surprised to discover that M9 and its immediate surroundings are cloaked in a very impressive double dark nebula. Most obvious is the oily

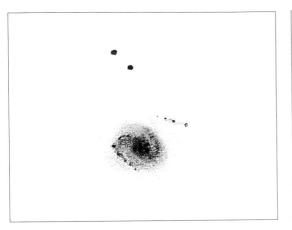

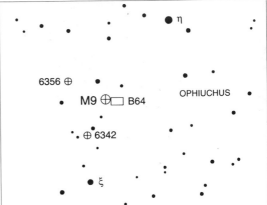

gulf known as Barnard 64 just to the cluster's west. A larger cloud, whose shape reminds me of a dinosaur footprint (with the toes pointing to the east), encompasses both Barnard 64 and M9. I wondered whether this gloomy apparition was the reason this globular looked so gray and dim. This may indeed be the case. M9 is one of the closest globulars to the center of our galaxy, and heavy absorption of the cluster's light by interstellar dust might dim its intensity by at least one magnitude.

Unfortunately, the cluster itself reveals little in the way of stunning detail; what's most conspicuous is what's absent. At low power it has a very strongly condensed center and a tiny unresolved halo, outside of which, to the south-southwest, lies a fine double star. Moderate power reveals some structure, namely, caps on the northwest and southeast edges studded with suns no brighter than 13th magnitude. Rosse clearly noticed the southern cap, seeing an outlying portion separated from the main body by a dark passageway. Thus the globular appears slightly oval with arms like those in a barred-spiral galaxy. High powers start to resolve the cluster, but only with difficulty.

Two smaller, fainter globular clusters lie just outside the boundaries of the dark envelope. NGC 6342 is a little more than a degree to the southeast, and NGC 6356 is situated equidistant to the northeast. This triangle of globulars – M9, NGC 6342, and NGC 6356 – "overlaps" a similar triangle of 6th-magnitude stars.

M10

NGC 6254
Type: Globular Cluster
Con: Ophiuchus
RA: 16^h 57^m.1
Dec: −4° 05′
Mag: 6.6
Dia: 19′
Dist: 14,300 l.y.
Disc: Messier, 1764

MESSIER: [Observed 29 May 1764] Nebula without a star, in the belt of Ophiuchus, close to the thirtieth star in this constellation, which is of magnitude six, according to Flamsteed. This is a beautiful, circular nebula; it can be seen only with difficulty with a simple three-foot refractor. M. Messier plotted it on the second chart of the path of the comet of 1769, *Mémoires de l'Académie 1775*, plate IX. Observed again 6 March 1781.

NGC: Remarkable globular, bright, very large, round; gradually brightening to a much brighter middle; well resolved with stars of 10th to 15th magnitude.

From dark skies, this 6.6-magnitude globular cluster in Ophiuchus is an easy binocular object and can be seen without too much difficulty with the unaided eye, especially because the 5th-magnitude star 30 Ophiuchi, an orange giant star of spectral type *K*4 III, is so close to it. Just stare at this star and your averted vision should pick up M10. To get to 30 Ophiuchi, I start at 2.6-magnitude Beta^1 (β^1) Scorpii (which has a blue, 5th-magnitude companion, β^2 Scorpii) and follow a line a little more than 10° (a fist-width) to the northeast to an equally bright and blue star, Zeta (ζ) Ophiuchi. Now continue on that line for another 6° (about two finger-widths) to a 5th-magnitude star, 23 Ophiuchi. Two and a half degrees farther along is 30 Ophiuchi. M10 is just 1° west-northwest of 30 Ophiuchi.

At low power M10 appears to be very typically detailed, but I could not escape the clear impression that its outer halo of stars has an ice-blue sheen, whereas its interior exudes a pale salmon light. Look for a yellow *spark* at the very center. These colors are similar to those of M3, though not as obvious. When I relax my gaze, I see two short, thin arms running

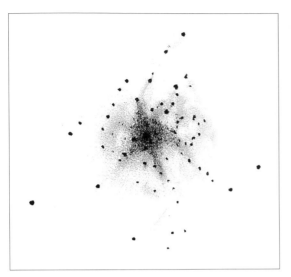

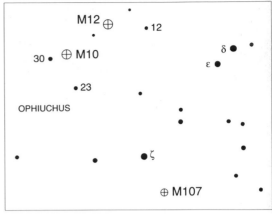

through the center of the cluster from northeast to southwest. Suddenly M10's core looks like Saturn with its rings seen nearly edge on through a thick fog.

Moderate power shatters this illusion and creates another. Now the cluster is caged – a ball of energy inside a pyramid of 12th-magnitude stars that may or may not belong to the cluster (they're at that discernible fringe in the 4-inch). This pyramid itself is enclosed in another, larger pyramid of stars. The cluster's center shines with a soft, uniform glow. When I use averted vision, the surrounding halo, which otherwise shimmers with a faint light, shoots sparks of starlight across that sheet of "plasma." This is not a flamboyant sight but an elegant one. Its starlight is rather demure and does not vary much in intensity. Most components look like diamond dust. Indeed, the cluster starts to resolve at magnitude 14.7, a tantalizing threshold for moderate apertures under dark skies. Why am I not surprised that only three variable stars have been discovered in this cluster? Even its sluggish recessional velocity of 43 miles per second speaks of a certain stately galactic reserve.

Although I have noticed several dark patches in the cluster's outer halo, I did not notice, as Rosse did, that the "upper" one-sixth of the cluster is fainter than the rest. Does such a variation in brightness exist? What do you see?

M11

Wild Duck Cluster
NGC 6705
Type: Open Cluster
Con: Scutum
RA: 18^h 51^m.1
Dec: –6° 16′.2
Mag: 5.8; 5.3 (O'Meara)
Dia: 13′
Dist: 5,460 l.y.
Disc: Gottfried Kirch, 1681

MESSIER: [Observed 30 May 1764] Cluster of a large number of faint stars, near the star *K* in Antinoüs [Aquila], which may be seen only with good instruments. With a simple three-foot refractor it resembles a comet. This cluster is suffused with faint luminosity. There is an eigth-magnitude star in the cluster. Kirch observed it in 1681, *Philosophical Transactions* no. 347, page 390. It is plotted in the large English *Atlas Céleste*.

NGC: Remarkable cluster, very bright, large, irregularly round, rich, one star of 9th magnitude among stars of 11th magnitude and fainter.

M11 is a small (13′) but extremely rich open cluster, bristling with the light of at least 680 suns; of these about 400 shine brighter than 14th magnitude. The cluster measures about 20 light years in diameter and its core is very dense. If you lived at the center of M11, you would see several hundred 1st-magnitude stars in the sky, and possibly 40 or so with an apparent brightness ranging from 3 to 50 times the light of Sirius, a scenario Robert J. Trumpler of Lick Observatory calculated in 1932.

To find M11, look not quite 2° southeast of 4th-magnitude Beta (β) Scuti. The cluster is visible to the naked eye as a fuzzy pellet of light flanked by two 6th-magnitude stars to its northwest. Through binoculars the view is almost as stunning as it is through a telescope. M11 sits in a notch in the north edge of the Great Scutum Star Cloud, the brightest stellar island outside the galactic center in Sagittarius. Barnard called this region the gem of the Milky Way.

At low power M11 is well resolved and displays a fluid fan of starlight streaming from an 8th-magnitude saffron star near the fan's apex. Although not a true cluster member, this brilliant sun is one of many yellow giant stars contained in this cluster, whose age is estimated to be some 500 million years.

When viewing the cluster at 23×, two scenes come to mind: an exploding party favor with countless flecks of confetti fleeing into space, or a volcanic vent blowing out incandescent gas and golden shards of molten rock. Can you see Smyth's "flight of wild ducks" (hence the nickname Wild Duck Cluster) in this V formation of stars? There is a pair of 9th-magnitude stars south of the eastern flock of ducks. Between this double

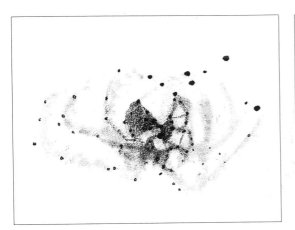

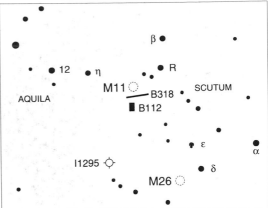

and the saffron star is a wall of faint stars. At moderate power, the cluster's nuclear region appears highly fractured, which was also noticed by the nineteenth-century astronomer Heinrich d'Arrest, who described M11 as "a magnificent pile of innumerable stars. Irregular and as if divided into several agglomerations."

Use high power to examine the cluster's dumbbell-shaped interior (see the accompanying drawing) and the waves of stars flowing away from it to the west. In fact, the outer fringes of the cluster are composed of wild stellar arcs that emanate from the cluster's northern extremes and curve southward. When I look with averted vision, I can see many faint streamers of stars at the limit of vision. With its triangular body, insectlike appendages, and tiny pointed northern "head," M11 looks more like a tick than a flock of wild ducks. (Does the tick's head look fuzzy to you?) If you can view the cluster with north at the top of your field of view, use high power and relax your gaze, or slightly defocus the cluster. I can make out a Jolly Roger (a grinning skull complete with a black eye patch).

If you have a very wide-field telescope you must take the time to sweep the Great Scutum Star Cloud, where there is an incredible diversity of bright and dark nebulosity (of various shades of gray). A striking slash of darkness (Barnard 318) lies immediately south of M11. South of B318 lies an eerily dark pond (Barnard 112). By the way, the notch in the northern edge of the Star Cloud (where M11 resides) is created by a massive swatch of darkness (Barnard 320).

One night in August 1995 when the sky was illuminated by the full moon I turned the world's second-largest refractor – the Lick Observatory 36-inch on Mount Hamilton in California – to M11 and studied it at 588×. The field of view was five times smaller than the cluster, so I had to electrically slew the telescope north and south, east and west, to see the entire cluster. While slewing west of the bright saffron star, a blizzard of faint starlight filled the field. Slewing further west, the view suddenly began to grow dim. I looked up to see if clouds had moved in, but the moon and stars were tack sharp against the velvet sky. Returning to the eyepiece I realized it was dark galactic vapors that had dimmed the starlight. I reversed the slewing direction and started over. This time I slowly "walked" across glistening moonlit fields, wet with starlight, until I gradually sank into that misty moor beyond them.

Just 1° northwest of M11 is a famous naked-eye variable, R Scuti, whose semiregular pulsations (from 4.8 to 6.0 magnitude) repeat about every 143 days. The star occasionally dips below naked-eye visibility.

Here's an interesting note: On 25 June 1995, while viewing M11 with 72×, I wrote in my notebook "very red star nearby." But I didn't indicate exactly where. Can you locate it? Is it a variable?

M12

Gumball Globular
NGC 6218
Type: Globular Cluster
Con: Ophiuchus
RA: 16^h 47^m.2
Dec: −1° 56′
Mag: 6.1; 6.8 (O'Meara)
Dia: 14′
Dist: 18,000 l.y.
Disc: Messier, 1764

The enormous summer constellation Ophiuchus harbors nearly two-dozen globular clusters within range of small telescopes, and seven of them were catalogued by Messier. Small but spectacular, M12 lies only 3½° northwest of M10 (a nearly identical globular) and 2½° east-northeast of 6th-magnitude 12 Ophiuchi. Use binoculars (or your telescope at low power) to view the clusters together for a twin treat. Separated by only

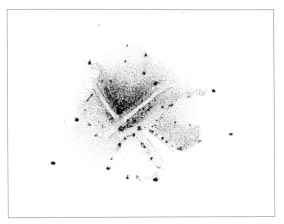

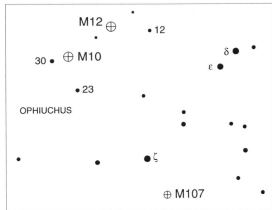

MESSIER: [Observed 30 May 1764] Nebula discovered in Serpens, between the arm and the left side of Ophiuchus. This nebula does not contain any stars, it is circular, and its light is faint. Near the nebula is a ninth-magnitude star. M. Messier plotted it on the second chart of the comet observed in 1769, *Mémoires de l'Académie 1775,* plate IX. Observed again 6 March 1781.

NGC: Very remarkable globular, very bright and large, irregularly round; gradually much brighter toward the middle; well resolved, stars of 10th magnitude and fainter.

3,700 light years, these two star-packed clusters are virtually cosmic neighbors. Each would appear as a roughly 4.5-magnitude object as seen by hypothetical inhabitants within the other. By comparison, Omega Centauri, the grandest globular visible from earth, shines at magnitude 3.9.

At 23×, there is a house- or rocket-shaped asterism just to the north of M12, and the globular looks like a puff of smoke from the rocket's exhaust, with its many bright members strung out like paper streamers in the wind. Smyth likened these linear features to a "cortège of bright stars," and Rosse saw them as long straggling tentacles.

Moderate power reveals a very loosely packed nuclear region surrounded by a faint halo of unresolved stars, though high power resolves the halo beautifully. I call M12 the "Gumball Globular," because that's what immediately came to mind when I first saw its wide assortment of bright and colorful stars.

Use 130× to appreciate the central star streams forming a wedge or triangle that fans to the south. A dark "V" borders it to the north. Otherwise, the star patterns favor the south and east. Three of M12's arms enclose dark bays. Can you spot them?

M13

Great Hercules Cluster
NGC 6205
Type: Globular Cluster
Con: Hercules
RA: 16^h 41^m.7
Dec: +36° 27'
Mag: 5.8; 5.3 (O'Meara)
Dia: 21'
Dist: 23,400 l.y.
Disc: Edmond Halley, 1714

MESSIER: [Observed 1 June 1764] Nebula without a star, discovered in the belt of Hercules. It is circular and bright; the center brighter than the edges, and is visible in a one-foot refractor. It is close to two stars, both of eighth magnitude, one above and the other below. The nebula's position has been determined relative to ε Herculis. M. Messier plotted it on the chart for the comet of 1779, which will be included in the Academy volume for that year. Seen by Halley in 1714. Observed again on 5 and 30 January 1781. It is plotted in the English *Atlas Céleste*.

NGC: Very remarkable globular cluster of stars, extremely bright, very rich, very gradually increasing to an extremely compressed middle, stars from 11th magnitude downward.

M13 is generally considered the finest globular cluster in the northern skies, mainly because it is visible to the naked eye in a well-known grouping of stars that sails high overhead in the summer sky. It is a swollen mass teeming with perhaps 300,000 to a half-million suns spread across 140 light years or more; a typical globular contains tens of thousands to hundreds of thousands of stars. A relatively close globular (about the same distance of M5), the Great Hercules Cluster is pleasingly bright. I determined its magnitude to be 5.3, which is slightly brighter than that recorded in most references. From dark skies and in good conditions, M13 is easily spotted as a fuzzy "star" with the naked eye, though it can be seen as a perceptible glow even through a light fog.

M13 lies about 2½° south of 3.5-magnitude Eta (η) Herculis, on the western side of the familiar keystone asterism in Hercules. At low power the cluster is bracketed nicely by two 7th-magnitude stars. The western one is a Type-*A*2 star, and the eastern one is a Type-*K*2 star, though, when I made my observations, both stars seemed to have a yellowish tint. Compare the color of the cluster's center with these stars and see if it doesn't appear yellow with a slightly greenish halo. John Herschel described the cluster as exhibiting "hairy-looking, curvilinear branches"; Rosse also noted the "singularly fringed appendages . . . branching out into the surrounding space." Two of these arms show prominently at 23×. They extend southeast and northwest from the nucleus and look like wings curving to the southwest. A forked "tail" of stars to the southwest completes this birdlike visage. Otherwise, the cluster at low power appears moderately condensed at the center with a gradual spreading out of light.

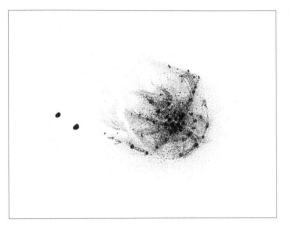

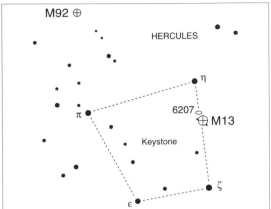

At moderate power this 14-billion-year-old cluster shows only a handful of seemingly bright members shining between 11th and 12th magnitude, though about two dozen 13th-magnitude stars sparkle into view across the cluster's face. The remainder form a barely resolvable haze, like finely crushed sandstone illuminated by a setting sun. Indeed, at times, I swear the center has a bloodlike tinge. This is not too difficult to believe, because the cluster's brightest members are red giants. At first, the core appears moderately diffuse, but if you stare long enough, you might see a gradual brightening toward the center.

M13 really packs a punch at high power! With averted vision the view is almost frightening – a blazing ball of tiny suns. So many more stars, so many more patterns to consider. Arcs of stars to the northeast create an impression of a strong galactic wind blowing from that direction, forming a bow shock on that side of M13. And the forked array of stars to the south-west forms a beautiful cock's-tail wake. Look carefully at the cluster's core; it is fractured, with a definite asymmetry toward the south. Here you can also see many dark patches and rifts. Only once did I recognize Rosse's classic dark Y shape, just southeast of the core, and, surprisingly, that was on a foggy night! I have yet to see it again in the 4-inch, despite having made several attempts to look solely for that feature. Interestingly, I discovered my own Y in the northwest halo, just inside the northern wing, as shown in the drawing. Furthermore, this dark Y shows nicely on a photographic plate made with the Lowell Observatory's 13-inch telescope.

M13 is an impressive cluster, but I think its grandeur is slightly overrated for *small*-telescope users. It certainly is a magnificent cluster in photographs and when seen through large-aperture telescopes. But is it the *finest* globular cluster in the northern skies for small apertures? The answer, in my opinion, is no. In a 4-inch refractor, M13 does not have as strong a visual impact as M5. To achieve truly outstanding resolution of M13 in a 4-inch telescope, one really needs high magnification and keen averted vision. The globular M5, on the other hand, immediately displays an assortment of stellar magnitudes, color, and star patterns at only 23×. Furthermore, M5 is, according to my magnitude estimate, only 0.4 magnitude fainter than M13. (However, *Sky Catalogue 2000.0* claims M5 is magnitude 5.75 and M13 is magnitude 5.86!)

Before leaving M13, don't forget to try for the 11th-magnitude spiral galaxy NGC 6207 about 40′ to the northeast. The Genesis at medium power shows it as an obvious elongated haze. These two objects provide a dramatic example of depth of field: NGC 6207 lies at a distance of 46 million light years – about 2,000 times farther than M13 in the halo of our own galaxy.

M14

NGC 6402
Type: Globular Cluster
Con: Ophiuchus
RA: 17^h 37^m.6
Dec: –3° 14′
Mag: 7.6
Dia: 11′
Dist: 33,300 l.y.
Disc: Messier, 1764

For the same reasons that M13 is popular, globular cluster M14 is not. It is beyond the normal naked-eye limit and lies in a region of sky belonging to an obscure constellation that never gets very high in the sky from mid-northern latitudes. Yet M14 is surprisingly detailed and deserves special attention.

To find it, first locate 3rd-magnitude Beta (β) Ophiuchi. Ten degrees (one fist-width) to the southwest is 4.5-magnitude 47 Ophiuchi, the brightest star in that region. M14 is just over 3° to the northeast and is visible in binoculars.

MESSIER: [Observed 1 June 1764] Nebula without a star, discovered in the drapery that hangs from the right arm of Ophiuchus, and lying on the parallel of ζ Serpentis. This nebula is not large, and its luminosity is feeble; however, it may be seen with a simple three-and-a-half-foot refractor; it is circular. Close to it is a faint star of the ninth magnitude. Its position has been determined relative to γ Ophiuchi, and M. Messier plotted its position on the chart for the comet of 1769, *Mémoires de l'Académie 1775,* plate IX. Observed again 22 March 1781.

NGC: Remarkable globular, bright, very large, round, extremely rich, very gradually becoming brighter toward its center, well resolved, 15th-magnitude stars.

At low power in the telescope this diminutive globular looks as lonely as it does in binoculars. The field is so sparse that the globular appears dim and distant in the void. This is another dusty region of the Milky Way, which contributes to light loss from the globular. But unlike M9, which looks gloomy, M14 shares none of that pallor. In fact, I find it quite luminous, like a comet with a diffuse inner coma surrounded by a diffuse outer coma. Overall the cluster has a pale straw color. Although there was not even a hint of resolution at low power, I noticed that what at first appeared to be a circular glow actually had a curved section of hazy light cutting north to south through the tiny, faint nuclear region. It also has a quite extensive halo, which was diminished greatly one night under high, thin clouds, so I wondered how city lights must affect this delicate object. Indeed, though M14 is catalogued as having a diameter of 11′, some popular references list a diameter as small as 3′.

The cluster starts to reveal itself at moderate power, when at first glance the center appears as a bulging mass, a cracking shell of starlight. Even the outer halo, which is is bracketed by two roughly 12th-magnitude stars, can be partially resolved. The cluster now bursts with faint starlight. That hazy north–south curve reveals patches of nicely resolved stars, like star-forming regions in the arms of a spiral galaxy. About seven distinct arms extend in various directions.

The cluster's center is loose enough to start resolving at moderate power, and how nice it is to see colorfully bright stars against the core. Higher power immediately produces a most interesting sight: a tiny stellar core emerges, tangerine in color. Perhaps it is the "faint sparkle" noted by Luginbuhl and Skiff in their *Observing Handbook and Catalogue of Deep-Sky Objects.* The rest of the nucleus is a jumble of bright chunks of starlight, whose north and south edges are slightly swollen, giving that region a slight dumbbell shape. Perhaps the spaces between these chunks

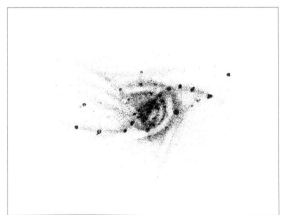

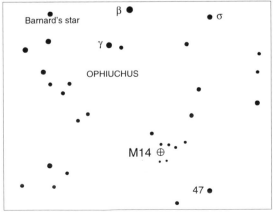

are what astrophotographer Isaac Roberts alluded to when he saw "vacancies in the centre" of the object on his glass plates.

The core is also surrounded by a ring of stars – a garland or rosette. Look closely, and you might see ripples of stellar haloes radiating from the nucleus to the southeast, as if a pebble were tossed at a sharp angle into that pond of stars. A dark lagoon to the southeast of the outermost halo is enclosed by a shoal of faint stars that connects to the southern arm. Finally, avert your gaze *way off* to one side, then relax. Do you see the nucleus burning as a lens-shaped mass of equally bright stars?

M15

Great Pegasus Cluster
NGC 7078
Type: Globular Cluster
Con: Pegasus
RA: 21^h 30^m.0
Dec: + 12° 10′
Mag: 6.3; 6.0 (O'Meara)
Dia: 18′
Dist: 30,600 l.y.
Disc: Jean-Dominique Maraldi II, 1746

MESSIER: [Observed 3 June 1764] Nebula without a star between the head of Pegasus and that of Equuleus. It is circular and the center is bright. Its position was determined relative to δ Equulei. M. Maraldi mentions this nebula in the *Mémoires de l'Académie 1746*. "I noted," he said, "between the stars ε Pegasi and β Equulei, a fairly bright, nebulous star, which consists of several stars. Its right ascension is 319° 27′ 6″, and its declination + 11° 2′ 22″."

NGC: Remarkable cluster, very large and bright, irregularly round, very suddenly much brighter in the middle, well resolved into very small [faint] stars.

Nearly a twin of M2 in Aquarius, this glittering gem in the winged horse, Pegasus, is one of six beautiful globulars brighter than 7th magnitude that grace the northern sky (the others are M2, M3, M5, M13, and M92). The Great Pegasus Cluster, M15, can be spotted without difficulty as a "fuzzy star" with the unaided eye, lying just 4° northwest of the topaz (Type-$K2$ I) 2nd-magnitude star Epsilon (ε) Pegasi. M15 is some 30,000 light years distant (16,600 light years farther away than M13) and measures up to 160 light years in diameter. Like M13, it contains many red-giant stars. But because of its greater distance, M15 appears fainter and more compact than M13.

At low power the cluster hides inside a triangle of three 7th- to 8th-magnitude stars. Hazy, spiderlike arms are already apparent, and the cluster brightens rapidly toward the center. Otherwise, like M13, most of M15's stars evade direct gaze and require averted vision. William Herschel rated this a good test object for resolution. But subtle details do show through. For example, it has a definite asymmetry. The late Harvard University astronomer Harlow Shapley first confirmed this by noting the oblateness at the cluster's central bulge, which is surrounded by a spherical shell of stars. M15 also displays dark patches. One obvious dark feature appears next to a detached string of stars on the northeast edge of the cluster's inner shell. Another, tighter arc of stars to the east of the nucleus makes the entire central region appear warped in that direction, something noticed by d'Arrest in the nineteenth century.

According to Webb, "Buffham, with a 9-inch [mirror] finds a dark patch near the middle with two faint, dark lines or rifts like those in M13." I did not notice these. My view is more like the one Isaac Roberts described at the turn of the century, in which stars are arranged "in curves, lines and patterns."

The cluster has an unusual resident, a 14th-magnitude planetary nebula, Pease 1, on its northeast side. In fact, M15 is one of two globular clusters known to contain a planetary; the other is a $10″ × 7″$ object, GJJC-1, in M22. F. G. Pease discovered the M15 planetary in 1928. But measuring only 1″ in diameter and awash in a sea of stars, Pease 1 is nearly impossible to detect in backyard telescopes. Its central star shines at magnitude 15.0. The nebula and the star eluded my gaze in the Genesis. The cluster is also host to a wealth of variable stars (nearly 100 are known). Jones points out that M15 ranks third behind M3 and the famous southern globular Omega Centauri in the number of variable stars it contains.

In 1974, M15 was discovered to be a source of x-ray energy, which, together with the cluster's apparent tightness, led some astronomers to conjecture that a black hole lurked at its center. The Hubble Space

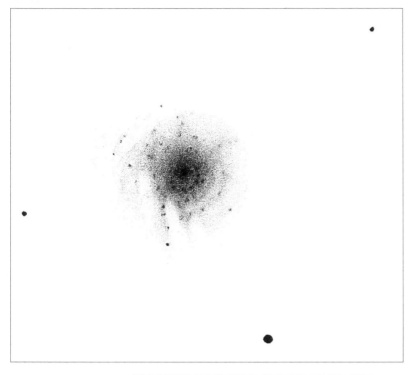

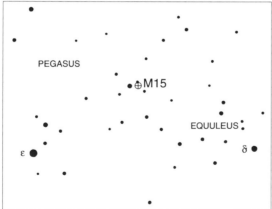

Telescope, however, disproved that by resolving the cluster virtually to its core and revealing nothing extraordinary about it. Instead, astronomers now believe that the x-ray energy might be coming from one or more supernova remnants.

M16

The Ghost, Eagle Nebula, or
Star Queen Nebula
NGC 6611
Type: Open Cluster and
Emission Nebula
Con: Serpens (Cauda)
RA: $18^h 18^m.8$ (cluster)
Dec: –13° 48′ (cluster)
Mag: 6.0 (cluster)
Dim: 120′ × 25′ (nebula)
Dia: 6′ (cluster)
Dist: 9,000 l.y.
Disc: Philippe Loys de
Chéseaux, 1746

MESSIER: [Observed 3 June
1764] Cluster of faint stars,
mingled with faint luminosity,
close to the tail of Serpens,
not far from the parallel of ζ in
that constellation. With a
small telescope this cluster
appears to be a nebula.

NGC: Cluster, at least 100
bright and faint stars.

Located 2½° west-northwest of 4.7-magnitude Gamma (γ) Scuti, in the spine of the Milky Way, is a most tantalizing sight – a fan of nebulosity (315 light years in extent, or about 20,000 times the diameter of the solar system) atop a loosely scattered cluster of stars. But as fine as M16 appears through a telescope, the view does not compare with the cornucopia of elaborate detail revealed in photographs made through large telescopes. Burnham, in his book, calls its photographic appearance "one of the great masterworks of the heavens," and his description of the nebulous opulence bears repeating:

> Thrusting boldly into the heart of the cloud rises a huge pinnacle
> like a cosmic mountain, the celestial throne of the *Star Queen*
> herself, wonderfully outlined in silhouette against the glowing fire-
> mist. . . . In the vast reaches of the Universe, modern telescopes
> reveal many vistas of unearthly beauty and wonder, but none,
> perhaps, which so perfectly evokes the very essence of celestial
> vastness and splendor, indefinable strangeness and mystery, the
> instinctive recognition of a vast cosmic drama being enacted, of a
> supreme masterwork of art being shown.

Visible as a hazy patch with the naked eye, the M16 complex occupies the northern end of a large S-shaped asterism of stars, which, if viewed together at 23× with north up, looks very much like a seahorse. In fact, the emission nebula itself, with north up, resembles a hand puppet, or a cartoonlike ghost flying through the night with outstretched arms and bulging eyes (thus my nickname the Ghost for the nebulosity).

At the center of the nebula, a stalagmite of blackness (20 trillion miles long) to the northeast, and a wedge of darkness to the north,

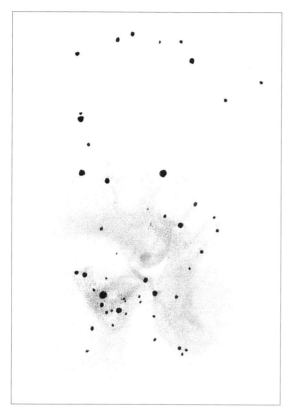

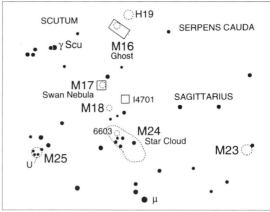

together create one of the most mystifying sights visible in galactic nebulae – tidal waves of dark matter that appear to be scrubbing away the bright coast of gas with their heavy ebb and flow. Earlier this century, astronomer Fred Hoyle believed that we were witnessing the expansion of hot gas into cooler gas, with the hot gas erupting like an exploding bomb. And that is largely what recent images from the Hubble Space Telescope have revealed: ebony pillars trillions of miles long dramatically being boiled away by the ultraviolet radiation of nearby stars. Shooting off the column tips are tapered nodules of gas (EGGs), each as wide as our solar system, where star formation is occurring.

The Ghost and its associated dark clouds are an extreme visual challenge in small telescopes, unlike the dark channels and swirls so clearly visible in M8. Do take up the challenge however. From Hawaii, the 4-inch can easily pick out the location of the dark northern wedge and reveal a definite V-shaped hole in the heart of the Ghost (the top of the Queen and her throne). A curious bright patch of nebulosity or unresolved clustering of stars lies just to the south of that V (see the drawing). In most pho-

The Messier Objects 77

tographs I have seen, this region is overexposed. If you spend the time carefully scrutinizing the southeastern extremities of the nebula, you might notice an absence of gas there – Burnham's "cosmic mountain." Can you trace out the faint wisps of dark nebulae between the mountain and the top of the throne?

The young, hot cluster illuminating the nebula dominates the northwest part of the Ghost. The cluster measures some 30 light years across and it is approximately 800,000 years old, though some of the youngest stars might be only 50,000 years old. There is a challenging double for the 4-inch near the center of the Ghost's head. It is just below its left "eye," in a faint stream of nebulosity.

Sharing the field of view with M16, and worth a look, is Harvard 19, a 12th-magnitude, cometlike open cluster 40′ to the northwest.

M17

Omega, Horseshoe, or *Swan Nebula*
NGC 6618
Type: Emission Nebula and Open Cluster
Con: Sagittarius
RA: 18^h 21^m.1
Dec: −16° 11′
Mag: 6.0 (cluster)
Dim: 40′ × 30′ (nebula)
Dia: 25′ (cluster)
Dist: 4,890 l.y.
Disc: Philippe Loys de Chéseaux, 1746

More Messier objects, namely 15, are located in Sagittarius than in any other constellation. And for good reason. The mythical Archer stands vigil in the direction of the center of our galaxy, the area most crowded with stars, dust, and gas. No wonder then that this parcel of sky yields the greatest variety and concentration of galactic star clusters and nebulae, including M17, which is a combination of both. With the exception of the Orion

Nebula (M42), M17 is the brightest galactic nebula visible to observers at mid-northern latitudes.

The Swan, as this emission nebula is often called, can be seen with the naked eye as a 6th-magnitude patch of light 2½° south of M16 and 2½° southwest of Gamma (γ) Scuti. In his 1889 *Celestial Handbook,* George F. Chambers was the first to compare this peculiar-shaped nebulosity to a swan floating on water. He was alluding to M17's brightest features, namely a long bar of gas (the swan's body), which is topped on the south-western end by a faint hook (the swan's curved neck). But with a glance at 23×, M17 first appears as a long blaze of gas and starlight slashed by dark lanes of obscuring matter; consider, now, that this bar spans 12 light years, or 72 trillion miles, of space. Camille Flammarion likened this lengthy feature to a "smoke-drift, fantastically wreathed by the wind," a wonderfully believable impression.

The faint hook of the swan's neck should materialize soon after you survey the bar. Stay with low power and let your eye drift across the field in all directions. The swan appears to be swimming in a faint mist rising from a black pool. With medium power, concentrate on the southern half of the swan and you might see long vapors rising off its back and neck. A prominent "check mark" of dark nebulosity forms the crook in the swan's neck (this is not to be confused with the bright nebula, which also has the appearance of a check mark). Now use high power to look at the star marking the western end of the bright bar. Immediately encompassing it are four bright knots of gas forming a Celtic cross.

The ghostly hook has given rise to the nebula's other nicknames – the Omega Nebula, because of its resemblance to the Greek letter "omega" (Ω), or the Horsehoe Nebula – names introduced by Smyth in the nineteenth century. Others have commented on the nebula's resemblance to the number 2. But, as my drawing shows, the 2, the Horseshoe, or the Swan is but a part of a vast and elaborate network of gas and dust. It takes a discerning eye and a combination of moderate and high powers to bring out the finest details within the nebulous regions.

For example, notice how the hook actually forms a complete loop, the northernmost portion being the most difficult to make out, requiring averted vision and patience. Note also the apparent absence of starlight within the loop. This is probably caused by a cloud of obscuring matter. Certainly this is the darkest region in the entire nebula; it looks like an ink stain in very long-exposure photographs. If you return to low power and really study the faint envelope of nebulosity surrounding the swan, which measures 40′ × 30′ (about half the size of the Orion Nebula), you will discover that it is not symmetric. It ends abruptly to the west of the swan's head and the "black hole." It's as if the swan has sailed westward across a

MESSIER: [Observed 3 June 1764] Streak of light without stars, five to six minutes long, spindle-shaped, and rather similar to that in the belt of Andromeda, but very faint. There are two telescopic stars nearby, lying parallel to the equator. Under a good sky, this nebula can be seen very clearly with a simple three-and-a-half-foot refractor. Observed again 22 March 1781.

NGC: Magnificent, bright, extremely large, extremely irregular shape, hooked like a "2."

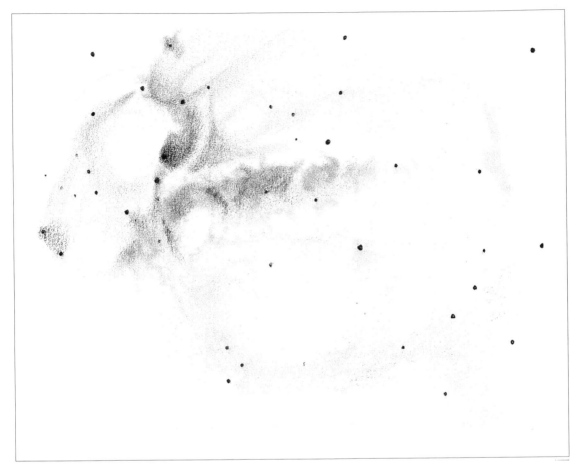

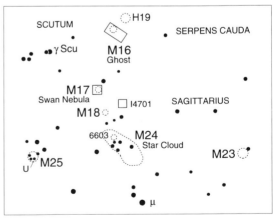

horseshoe-shaped pond to a shore of black sand. Curiously, about one swan diameter to the northwest is the tiny glow of emission nebula IC 4706. Could these glows be related, being separated only visually by a swath of foreground dark nebulosity? If so, you can imagine the swan looking across this dark gulf at its isolated cygnet.

Unlike the obvious star clusters found within M8 and M16, the one inside M17 appears to be nothing more than a weak cast of 35 stars of 9th magnitude or fainter. But, in fact, some 660 suns are sprinkled across the tortuous confines of this gaseous nebula.

M18

Black Swan
NGC 6613
Type: Open Cluster
Con: Sagittarius
RA: 18^h 20^m.0
Dec: −17° 05′.9
Mag: 6.9
Dia: 5′
Dist: 4,000 l.y.
Disc: Messier, 1764

MESSIER: [Observed 3 June 1764] Cluster of faint stars, slightly below the previous one, number 17, surrounded by faint nebulosity. This cluster is less obvious than the penultimate one, number 16. With a simple three-and-a-half-foot refractor, this cluster looks like a nebula, but with a good telescope only stars are visible.

NGC: Cluster, poor, very little compressed.

M18 lies only 1° south of M17, near the extreme northern edge of the Small Sagittarius Star Cloud (M24). Burnham calls this loose, 5′-wide gathering of stars (covering only 6 light years of space) one of the most neglected Messier objects, which is sometimes omitted from lists of galactic star clusters. Although credited with having only 40 members, the cluster is surrounded by faint background stars that add to the visual pleasure.

At 23×, M18 shares the field with M17, and the two are separated by a large but faint S-shaped string of similarly bright stars. I call M18 the Black Swan, because the main body of bright stars forms a pattern reminiscent of the bright nebulosity in M17, but unlike the rather sleepy looking M17 Swan, the M18 Swan is raising one of its large wings – a black wing (a region devoid of bright stars) outlined by five roughly 10th-magnitude stars. Of course, Black Swan is a double entendre, because it also refers to the fact that this cluster is often ignored or neglected.

Contrary to what is sometimes stated – that M18 is best viewed at low

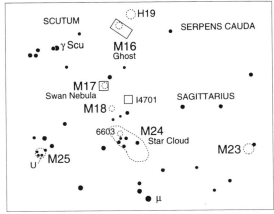

power, when it appears small and concentrated – I find the cluster comes to life at 130×, when the faint background stars of this rich Milky Way region enhance the view of the otherwise subtle grouping of 10th- to 11th-magnitude stars. A nice double star within it also adds a bit of sparkle. Admittedly, low power places M18 in a very favorable light: with the big Swan (M17) to the north, a rich swath of nebulosity (IC 4701) to the northwest, and the Small Sagittarius Star Cloud (M24) to the south. This tiny, seemingly insignificant cloud of meager starlight is surrounded by dazzling cosmic giants.

Although I did not notice any nebulosity visually, photographic plates made with the 48-inch Schmidt telescope on Palomar Mountain in California reveal that the cluster is bathed in a faint nebulous glow. I wonder what size telescope is required to see this? Meanwhile, can you make out the wishbone pattern of stars to the southwest (the swan's head) and the dim stream of 12th- to 13th-magnitude stars outlining the southern tip of the upraised wing?

M19

NGC 6273
Type: Globular Cluster
Con: Ophiuchus
RA: 17^h 02^m.6
Dec: −26° 16′
Mag: 6.8
Dia: 14′
Dist: 34,500 l.y.
Disc: Messier, 1764

MESSIER: [Observed 5 June 1764] Nebula without stars, on the same parallel as Antares, between Scorpius and the right foot of Ophiuchus. This nebula is circular; it is clearly visible with a simple three-and-a-half-foot refractor. The known star closest to this nebula is sixth-magnitude Flamsteed 28 Ophiuchi. Observed again 22 March 1781.

NGC: Globular, very bright, large, round, very compressed in the middle, well resolved. It consists of stars of 16th magnitude and fainter.

Despite what the *NGC*'s description says, M19 in Ophiuchus is a challenging object to resolve. Although the cluster shines with a total magnitude of 6.8, the average brightness of its most luminous suns is about magnitude 14. From dark skies, you can spot M19 with the naked eye 3° west of 4th-magnitude 36 Ophiuchi and ½° south and slightly west of 7th-magnitude 28 Ophiuchi. M19 is one of the most elongated globular clusters known; Shapley estimated that the cluster contained twice as many stars in its major axis as in its minor axis. Even a glimpse through the 4-inch telescope at low power reveals this ellipticity, though the telescope probably only reveals half of the 140-light-year-wide orb.

At 72×, there is not much difference: an unresolved haze gradually diffuses out from a bright stellar nucleus. High power shows about a half-dozen stars, most of which hug the outer fringe of the cluster's halo at the main cardinal directions. M19 is oblate north to south, though the multitude of unresolved stars surrounding the cluster's core favor the west. (Jones saw the cluster as being 10 to 15 percent longer north to south than east to west.) Noteworthy are some spiral-like arms of stars, which in the south appear to curve counterclockwise, whereas those in the north curve clockwise.

The globular is also colorful: I see a topaz core surrounded by swirls of blue smoke. Smyth found the stars to be of creamy white tinge, and slightly lustrous in the cluster's center. Do you see the numerous dark patches that stain the cluster, making it appear mud-splattered? I find these patches particularly intriguing, because the cluster resides in a bright gulf of starlight surrounded by the vast naked-eye rivers of darkness

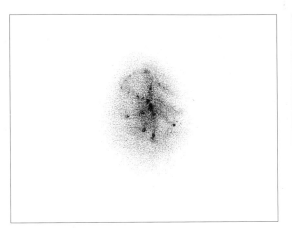

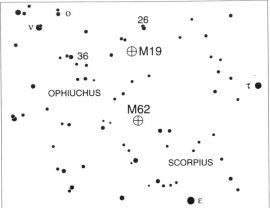

that cut through the Scorpius Milky Way region. Could these stains be tiny black clouds silhouetted against the more distant globular? It's a question to consider, because this globular belongs to a group of such clusters concentrated around the galactic center (M19 lies a little more than 3,000 light years from it). Light traveling from these distant objects is dimmed by intervening dust and gas, so it's hard to judge their sizes; foreground stars in the rich Milky Way can easily appear disguised as globular cluster members. The question, then, is whether M19 is truly elliptical, or is this an illusion created by interstellar absorption?

M20

Trifid Nebula or *The Clover*
NGC 6514
Type: Nebula and Cluster
Con: Sagittarius
RA: 18^h 02^m.5
Dec: −23° 02′
Mag: 6.3 (cluster)
Dim: 20′ × 20′ (nebula)
Dia: 28′ (cluster)
Dist: 5,000 l.y.
Disc: Guillaume Le Gentil, probably 1747

The 2°-wide expanse of Milky Way encompassing M8, M20, and M21 is the most dramatic Messier field in the entire sky. At 23×, these objects and several other clusters and nebulous patches (both bright and dark) fill the field. Such a tight gathering of nebulous splendor might be nothing more than a chance alignment. M8 and M20 are possibly part of the same complex; certainly their distances (5,200 and 5,000 light years, respectively) suggest they could be. And even though M20 and M21 are separated by about a thousand light years, I cannot discuss one without also discussing the other, because they look like they belong to the same celestial microcosm.

If you look at this region – which is about 5° (two finger-widths) west-northwest of 3rd-magnitude Lambda (λ) Sagittarii – with the naked eye, you should immediately see two hazes (M8 and M20) making an arc with the 5th-magnitude star 4 Sagittarii ½° to the west. Use binoculars or a wide field of view to see the "fishhook" of Milky Way (NGC 6526) between M20 and M8. Now concentrate on the interior of the fishhook. Do you see a "black hole"?

MESSIER: [Observed 5 June 1764] Star cluster, slightly above the ecliptic, between the bow of Sagittarius and the right foot of Ophiuchus. Observed again 22 March 1781.

NGC: A magnificent object, very large and bright, trifid, a double star involved.

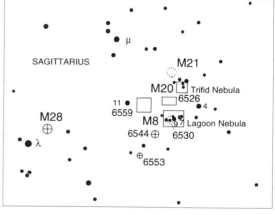

M21

NGC 6531
Type: Open Cluster
Con: Sagittarius
RA: 18^h 04^m.2
Dec: –22° 30′
Mag: 5.9
Dia: 15′
Dist: 4,100 l.y.
Disc: Messier, 1764

MESSIER: [Observed 5 June 1761] Star cluster close to the previous one [M20]. The known star closest to these two clusters is seventh-magnitude Flamsteed 11 Sagittarii. The stars in these two clusters are of eighth and ninth magnitude, and are surrounded by nebulosity.

NGC: Cluster, pretty rich, little compressed, stars from magnitude 9 to 12.

My guess is that Messier and his contemporaries must have first noticed these hazes with their naked eyes (as they probably did with M24, M42, M44, and M45, etc.). But when Messier turned his telescopes on M20, he noticed not a nebula, but a cluster of stars. Only in his description of M21 does he say that both M21 and M20 are enveloped in nebulosity. Messier never resolved the tiny glow of the Trifid in his small telescope, and the cluster he referred to is in the southern portion of the "cruciform group" cited by Webb.

Webb's Cross is a fine collection of roughly a half-dozen 6th- and 7th-magnitude stars (shaped like a cross with warped arms) that appears to the naked eye as a single nebulous patch; M21 marks the northern tip of the cross and M20 the base. So I can understand Messier's confusion: to the naked eye, these two clusters appear to be immersed in a cocoon of galactic gas, which vanished when he employed his optically inferior telescope. By the way, like the crossbow of stars in M8, I prefer to recognize Webb's Cross as an asterism, though I wonder if it is really a cluster.

There is a minor mystery here. Messier notes that the brightest star in M21 is 11 Sagittarii, but that star lies 2° to the southeast of the presently recognized cluster! He obviously misidentified the star, but someone who likes to model history from shreds of evidence might have fun trying to decipher exactly what happened here. I precessed Messier's 1764 positions for M20 and M21, and everything seems to be in order.

John Herschel is credited with being the first to call M20 the Trifid Nebula. A small but noble cloud of glowing gas, M20 looks more like a four-leaf clover glazed by frost than a tri-lobed nebula. The name "trifid" is

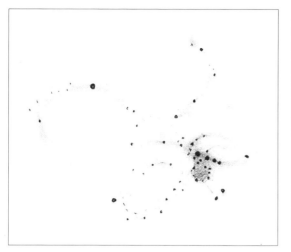

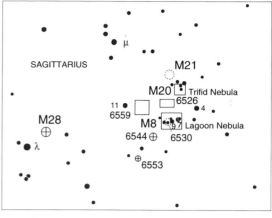

very deceiving (especially to anyone who first sees it in a photograph); it refers only to the nebula's three brightest southern portions. Do you find it strange that John Herschel would have called this nebula a trifid when his father, William, had already catalogued it in *four* portions? Anyway, I prefer the clover metaphor for two reasons. First, M20 has the shape of one, and second, because the fourth lobe is faint in small telescopes, you should feel lucky if you glimpse it!

The leaves of the clover are separated by dark lanes of obscuring matter emanating from a central well of darkness. Use high magnification to peer into the well and, given stable atmospheric conditions, you might see the discrete triple star HN 40. The two brightest members (7.6 and 8.7 magnitude) display a striking color difference. The brighter one, a Type-O7 star, for some reason has a mustard hue, while the other is a dying ember, charcoal-colored with a spark of red. The third star shines at magnitude 10.7 and looks colorless in the 4-inch. The 7.6-magnitude primary of HN 40 appears to be M20's main source of illumination, though other hot stars, cloaked by dark matter, could contribute energy as well.

I do see a trifid nebula at low power, though it is not the one Herschel refers to. The Trifid itself is one leaf of that; a ball of gas surrounds the 7.4-magnitude star just 10′ to the north (the second leaf). Now, use averted vision to follow the streamers of celestial fog that waft to the west. Do you see where more gas is illuminated by the 6th-magnitude star marking the western arm of Webb's Cross? All the nebulosity associated with the Trifid spans about 30 light years in space, and its total apparent size is two-thirds that of the full moon.

Less than a degree northeast of the Clover glimmers M21, a bright spread of young stars, about 50 of which are easily visible in small telescopes. Although it is listed as covering an area of 15′, M21's boundaries are poorly defined. Use high power to pierce the hazy heart of M21, the haze being an illusion created by numerous unresolved stars. Do you see the cluster's "spiral" structure? I can follow arms of stars flowing away from the central triangle of stars. One of these stars looks blue with a hint of yellow, while another looks blue with a hint of red. Overall, the cluster's stars appear white with just tinges of color. Use low power to appreciate the linear alignment of many of its fainter members. With a little flight of fancy, I see the hazy heart with its flashes of star color as a firefly, and the swirls of stars as its whimsical path across the heavens.

Here is a naked-eye challenge for you. Can you resolve Webb's Cross and separate M21 from M20?

M22

Great Sagittarius Cluster or
Crackerjack Cluster
NGC 6656
Type: Globular Cluster
Con: Sagittarius
RA: 18^h 36^m.4
Dec: −23° 54′
Mag: 5.2
Dia: 33′
Dist: 10,100 l.y.
Disc: Abraham Ihle, 1665; John Hevelius apparently noticed it before then

Without question, M22 should be called the "Great Sagittarius Cluster." It is a bonfire of a half-million suns that blazes at magnitude 5.2 and measures 33′ across – about the apparent diameter of the full moon. It ranks third only to Omega Centauri and 47 Tucanae among globulars in brightness and apparent size. Burnham writes that J. R. R. Tolkien penned an exquisite description of M22 in *The Hobbit* when he spoke of the fabulous jewel called the Arkenstone of Thrain: "It was as if a globe had been filled with moonlight and hung before them in a net woven of the glint of frosty stars."

No matter what size telescope I have aimed at this globular – from the 4-inch Genesis to the 36-inch Clark refractor at Lick Observatory – I am always awed by its grandeur. Ironically, the Tolkien quote describes dead-on the view through the Lick 36-inch at 588×. Although the full moonlight that August evening in 1995 suppressed the cluster's overall brilliance, its several hundred thousand suns still glistened with a faint silvery sheen, and each star looked like frost clinging to a frozen window-pane. On moonless nights, few sights compare to M22's tempest of starlight projected against the Milky Way.

At a distance of only 10,100 light years, M22 is among the closest globular clusters – closer than any visible in the northern sky. Its location 9° south of the galactic plane, however, diminishes its true radiance because of the intervening dust. To the naked eye it is a tight bead of light, lying about 2½° northeast of the *K*2 III star Lambda (λ) Sagittarii. At low power M22 appears more oval than round, with the major axis running slightly east of north. Let your eye relax, and with time you will notice that the oval glow is surrounded by a faint halo of unresolved stars that has prominent extensions to the north and south. A powerful orange star punctuates the northern arm.

My pencil drawing shows the view mainly with high power, so it does not capture the cluster's many hazy extensions and its enormous halo. I call M22 the Crackerjack Cluster (after the sweet popcorn treat with the prize in every box) because, at high power, after you penetrate the outer

MESSIER: [Observed 5 June 1764] Nebula, below the ecliptic, between the head and bow of Sagittarius, close to the seventh-magnitude star Flamsteed 25 Sagittarii. This nebula is circular, does not contain any stars, and is clearly visible in a simple three-and-a-half-foot refractor. The star λ Sagittarii was used to determine its position. Abraham Ihle, the German, discovered it in 1665 when observing Saturn. M. le Gentil observed it in 1747, and published a drawing, *Mémoires de l'Académie 1759*, page 470. Observed again 22 March 1781. It is plotted in the English *Atlas Celéste*.

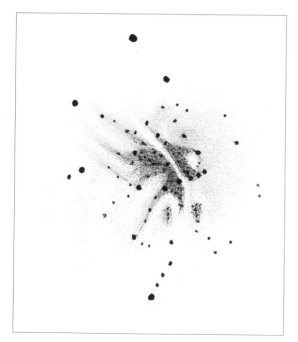

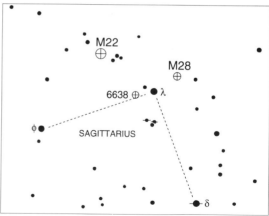

NGC: Very remarkable globular, very bright and large, round, very rich and much compressed, stars from 11th to 15th magnitude.

swarms of stars, many surprises await your gaze. The core of the cluster appears to sit in a hook of 10th- and 11th-magnitude stars on the south and west sides. It looks as if the cluster has been snagged by this hook of stars while drifting through space, like flotsam in a river getting snared by a hanging branch. And, as Skiff logs, a prominent clump of five stars – a fuzzy cluster within the cluster – can be found in the northeast quadrant of the core. Most striking, however, is the dark gash running southwest to northeast across the core. It is the most prominent rift in any globular I have seen, and it cleanly splits the nuclear region in half! Furthermore, dark bays run into the southern part of the nucleus from the southeast and two stellar deltas seem to have formed in the mouth of the dark river to the south.

To view the inner details most effectively, I observe the cluster in twilight, when the outer distraction of stars fades into the emerging dawn. This visual trick works for other bright globulars as well.

Like the globular cluster M15, M22 contains a planetary nebula, GJJC-1 (which has been identified as the infrared source IRAS 18333-2357). Its tiny disk measures $10'' \times 7''$ and its central star shines at magnitude 14.3. Proper motion studies have shown the planetary to be a part of the cluster.

M23

NGC 6494
Type: Open Cluster
Con: Sagittarius
RA: $17^h 56^m.9$
Dec: $-19° 01'$
Mag: 5.5
Dia: 30'
Dist: 2,100 l.y.
Disc: Messier, 1764

The Sagittarius Milky Way contains so many notable Messier objects – the Lagoon Nebula (M8), the Trifid Nebula (M20), the Swan Nebula (M17), the Great Sagittarius Cluster (M22) – that it is easy to overlook some of the smaller treats. M23, for example, all but hides in the southwest corner of a dark and inconspicuous valley that slices through the hub of our galaxy. Yet this open cluster of 150 or so stars spread across 30′ (one full moon diameter) is deceivingly dynamic. Train your binoculars on this region, about 5° due west of the Small Sagittarius Star Cloud (M24) or 4½° north-west of 4th-magnitude Mu (μ) Sagittarii, and see if the cluster doesn't remind you of a spider lying in wait for its prey. That M23 lies at the center

MESSIER: [Observed 20 June 1764] Star cluster between the tip of the bow of Sagittarius and the right foot of Ophiuchus, very close to the star Flamsteed 65 Ophiuchi. The stars in this cluster are very close to one another. Its position was determined relative to μ Sagittarii.

NGC: Cluster, bright, very large, pretty rich, little compressed, stars of 10th magnitude and fainter.

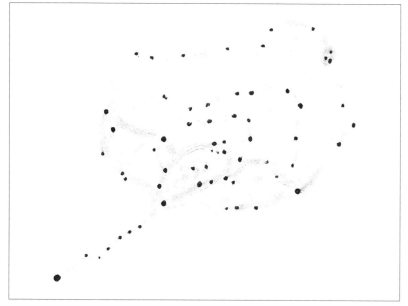

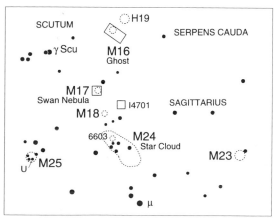

of an apparent hole in the Milky Way might also lead one to think that it has devoured the stars around it.

The arcing patterns of the cluster's 75 brightest stars entice the telescopic viewer to form a creative mental image of some kind. Burnham suggests an outline of a Chinese temple, or a bit of calligraphy. John Mallas saw a bat in flight, and I couldn't agree more. Admittedly, I am partial to bats, so I need no further convincing.

You really need low power and a north-up orientation of the cluster to fully appreciate the bat motif (see the drawing). A faint string of stars flowing southeast of an 8th-magnitude star meet with a decidedly elliptical pattern of stars to form the bat's tail and body. Its wings arc to the southwest and northeast. Opposite-curving arcs southeast of the stellar ellipse seem to flow away from the head of the bat. With a stretch of the imagination, I see these waves of starlight as the bat's ultrasonic screams. A tight triangle of stars in front of the bat could be the insect it has echolocated. Can you see the dark stream of nebulosity between the bat's head and the first ultrasonic wave?

Creative perception is an important part of astronomy's heritage. What are the constellations but figments of some creative individuals' minds. Some familiar star patterns, like Orion's Belt and the Big Dipper, are even household names. Looking for shapes, likenesses, and fanciful imagery in patterns of stars and nebulosity makes stargazing more fun and accessible, and easier to share with others. It can also help you remember subtle details of an object or region of sky. Memorizing star patterns through the telescope can be the first step to making discoveries. If, for example, a new star (a nova) were to appear near M23 – disrupting the square, or the boat, or the bat motif you had become familiar with – you would recognize that something was different; the shape you created would appear altered. This is how nova hunters George Alcock and Peter Collins each have discovered four new stars – using binoculars!

At 72× the bat loses its impact. Try, however, to locate what looks like a little spoon holding a peanut (again, view with north up). At high power, the entire cluster is clumped in little isolated groups of stars separated by meandering rivers of dark nebulosity, which seem to empty into two pools of darkness (on either side of the bat's tail). Using 130×, see if you can locate a trio of double stars in the cluster's northwest quadrant.

Return to low power and explore the Milky Way around M23, especially to the west, where you'll find a great crack in the stellar expanse – the cosmic cave from which the bat emerged.

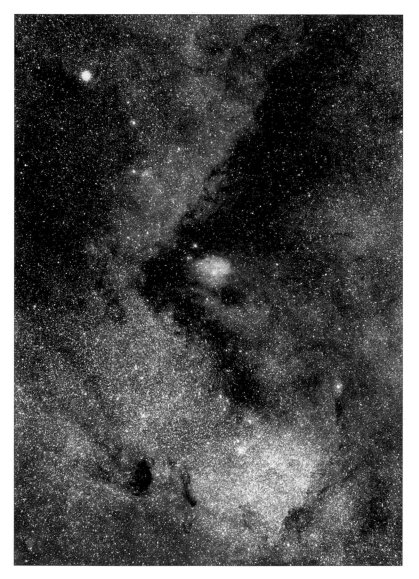

M24

Small Sagittarius Star Cloud
Type: Star Cloud (M24) and
Open Cluster (NGC 6603)
Con: Sagittarius
RA: 18ʰ 17ᵐ.4
Dec: −18° 36′
Mag: 4.6 (star cloud); 2.5
(O'Meara)
Dim: 1° × 2° (star cloud)
Dia: 7′ (cluster)
Dist: 9,400 l.y.
Disc: Messier, 1764

MESSIER: [Observed 20 June 1764] Cluster on the same parallel as the previous one [M23], and above the tip of the bow of Sagittarius, in the Milky Way. A large nebula, within which there are several stars of different magnitudes. The luminosity that is spread throughout the cluster is concentrated in several regions. The position given is that of the center of the cluster.

NGC: Remarkable cluster, very rich and very much compressed, round, stars of [12th] magnitude and fainter, in the Milky Way. [The *NGC* description is of NGC 6603.]

Between the Lagoon Nebula (M8) and the Swan Nebula (M17) lies one of the most impressive stellar cities visible in small telescopes. M24 is not a true galactic cluster, but a rectangular-shaped star cloud measuring 2° by 1°. Of all the Messier objects, M24 is second only to the Andromeda Galaxy, M31 (3° × 1°), in apparent size. Commonly called the Small Sagittarius Star Cloud, M24 is a virtual carpet of stellar jewels, laid out across 330 light years of space. To the eye, it is so big that estimating its brightness is tricky: you have to defocus comparison stars until they've ballooned to four moon diameters, so that the comparison star's light is

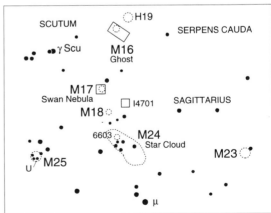

spread over an area of sky equal to that of M24. Most references list the star cloud's magnitude as 4.5, but this is too faint; I place it a full two magnitudes brighter, at 2.5. Because its light is spread out over such a large area, its surface brightness is low, making the cloud appear dimmer than it really is. Light pollution has robbed many of us of the privilege of enjoying this summertime wonder. If you are a city dweller plan to spend some time admiring this galactic treasure under a truly dark-sky site.

Through the Genesis at 23×, no sight in the visible universe shares M24's mystical qualities. I was amazed to find this seemingly three-dimensional patch of Milky Way beaming with a distinctive pale green sheen, suggesting a composition more of gelatin than starlight, a giant Euglena wrapped in stellar filigree.

The entire star cloud is blotted with pools of dark nebulosity. One prominent pool (Barnard 92) resides on the cluster's northwest side; it is surrounded by a succession of moderately bright stars. Use high power to concentrate on the center of the pool. Do you see your "reflection" – an illusion created by the light of a solitary 12th-magnitude star shimmering in the center? A wide canal of darkness to the north of M24 creates a sharp northern border. But the cloud seems to continue beyond the canal in a tapering spit toward the Swan Nebula.

Equally remarkable are the vast tracts of dark nebulosity to the west and east. These rival the star cloud in splendor. When I place M24 in the center of the field of view at 23×, then move one field to the northwest, I feel as if I've stepped off a steep cliff of starlight and am free-falling into a dizzying celestial abyss. If you relax your gaze you might see the ever-widening wedge-shaped array of dark veins running off to the southeast. Wide-field photographs of this region reveal a dark corridor about the size of M24 to the southeast. It looks as if someone has lifted M24 like a log from that spot and dropped it in its present location, leaving behind its barren imprint in the celestial "soil."

Notice the description of M24 in the *NGC,* which obviously could not have been referring to the same object as Messier's "large nebulosity" with a diameter of 2°. M24 was formerly misidentified as the 11.4-magnitude star cluster NGC 6603, which resides in the northern portion of M24, just 15' north-northeast of an orange 6.4-magnitude *K*5 star. Messier probably would not have seen this tiny, compact glow, which shines feebly with the light of an out-of-focus 11th-magnitude star. Even from the dark skies of Hawaii, I nearly overlooked that minuscule fuzzy knot, which at low power affords only a hint of resolution. High power reveals a semicircular dark patch just south of it. Whereas for most objects I recommend a long, steady gaze to distinguish a particular feature, to see this dark loop you must give a quick, brief glance. That's because for dark nebulae, the longer you look in the eyepiece, the more dark nebulosity you see, and the dark feature you're interested in soon gets lost in a complex web of dark streamers and spots that emerge from the background, especially with averted vision. In fact, if you spend a few minutes staring at M24, the star cloud will suddenly be swarming with dark veins.

Finally, compare the color of M24 with that of a smaller star cloud south of Mu (μ) Sagittarii. The Mu Sagittarii cloud looks gray and diffuse; I once thought I had breathed on the lens. It contrasts with the greenish glow of M24!

M25

IC 4725
Type: Open Cluster
Con: Sagittarius
RA: 18^h 31^m.7
Dec: –19° 07′.2
Mag: 4.6
Dia: 30′
Dist: 2,300 l.y.
Disc: Philippe Loys de
Chéseaux, 1746

MESSIER: [Observed 20 June 1764] Cluster of faint stars near the two preceding ones [M23 and M24], between the head and the tip of the bow of Sagittarius. The known star closest to this cluster is sixth-magnitude Flamsteed 21. The stars in this cluster are difficult to see with a simple three-foot refractor. No nebulosity is visible. Its position has been determined relative to μ Sagittarii.

IC: Cluster, pretty compressed.

Smyth, in his *Cycle of Celestial Objects,* captured the immediate impression of M25: "a loose cluster of large and small stars. . . . The gathering portion of the group assumes an arched form, and is thickly strewn in the south, or upper part, where a pretty knot of minute glimmers occupies the centre, with much star-dust around." This bright open cluster in

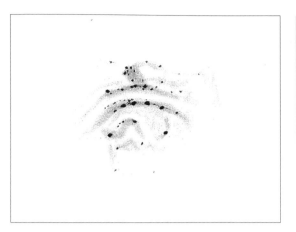

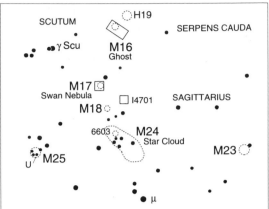

Sagittarius makes an easy naked-eye target and is finely resolved in small telescopes. It lies about 3½° east-southeast of the Small Sagittarius Star Cloud (M24) and just south of an inverted spoon of binocular stars. Its estimated 600 stars stretch across 20 light years of space, and the 30 or so bright ones that we can see cover a ½° area of sky. M25 is the only Messier object cross-referenced in the *Index Catalogue* (IC).

The main body of stars forms a north–south oriented figure eight with legs extending from all sides; in this way the cluster looks very beetle-like. The western half of the southern section of the "8" is filled with stars; this curious-looking arrangement is Smyth's pretty knot of minute glimmers. Just east of the knot blazes the Cepheid-type variable star U Sagittarii, a member of M25, whose magnitude waxes and wanes between 6.3 and 7.1 about every 6 days, 18 hours; near minimum the star's light becomes yellower. With a little effort you can chart its subtle variation in intensity and derive at its period. Can you see this star with the naked eye when it is at maximum?

At medium power several strong arcs of starlight spread across the field of view like rippled dunes glistening in a midday sun. Shadows fill the troughs between each crest and fade into the misty light of the stellar background. (On the off chance that you have had the opportunity to study an alligator's foot up close, the tapered digits of these star streams will look surprisingly familiar.)

M25's stellar arcs are so visually stunning that the cluster is beautiful for that reason alone. Once again, we have the opportunity to focus attention not on the bright stars but on the wisps of dark nebulosity – it's a skill worth developing, because it will add a new dimension to your observing. First use low power, and let your eye drift until it finds a comfortable position for averted vision. Then relax your gaze; M25's dark central stream becomes inky black and seems to lengthen well beyond the eastern and western edges of the cluster. The rest of the cluster looks like a child has taken a felt-tip pen to it and begun inking out the stars row by row.

After studying M25 look immediately to its east. There you will see a patch of Milky Way about the same angular size of M25 that looks like an afterimage of the cluster.

M26

NGC 6694
Type: Open Cluster
Con: Scutum
RA: 18^h 45^m.2
Dec: –9° 23'.1
Mag: 8.0
Dia: 8'
Dist: 5,000 l.y.
Disc: Probably Guillaume Le
Gentil, before 1750

MESSIER: [Observed 20 June 1764] Cluster of stars close to the stars *n* and *o* of Antinous [now ε and δ Scuti], and among which there is one that is brighter. With a three-foot refractor they cannot be detected; a good instrument must be used. This cluster contains no nebulosity.

NGC: Cluster, quite large, pretty rich and compressed, stars from 12th to 15th magnitude.

I first encountered M26 during a comet sweep in the rich Milky Way region in and around Scutum – perhaps just as Messier had done more than 200 years before. Using 23×, I noticed the cluster's concentrated glow enter the eyepiece, but it looked so small that I paid it little attention, figuring it was just an asterism; more intriguing was the 8th-magnitude globular cluster NGC 6712 about 2° to the northeast. After admiring the globular, a twinge of curiosity caused me to return to that tiny asterism, in which an ice-blue gem sparkled conspicuously amid a sprinkling of diamond dust. Suddenly I realized that this speckled glow was M26. From brighter skies, this object might well be overlooked, because its tantalizing gleam of background stars would be all but washed out. M26 is an innocuous knot of stars which spans an area of 12 light years and contains 120 stars in a disk 8′ in apparent diameter, though only about two dozen of the suns are readily seen in the 4-inch. That's pretty condensed compared to other Messier open clusters. M25, for example, is about four times larger in apparent diameter and measures 20 light years across, though it is also twice as close as M26. The brighter of M26's stars shine at only magnitude 10.3, for a combined magnitude of 8. You'll find the cluster less than 1° east-southeast of 5th-magnitude Delta (δ) Scuti and about 3½° south-southwest of M11 in the stunning Scutum Star Cloud.

The entire region surrounding M26 appears mired in dark nebulosity and the low-power field is haunted by fleeting hazes of unresolved stars. Moderate power shows two rows of stars extending to the north from the cluster's central diamond of stars. These pincers are clearly divided by a lane of obscuring matter, though other observers have documented it as a "hole," bereft of stars. James Cuffey of Indiana University first noticed the black heart of M26 in 1940, estimating that the star density at the cluster's core was 13 percent less than in the regions immediately surrounding this 3′-diameter zone of darkness. The diamond is further caged to the southwest and northeast by two thicker bars of blackness. With averted vision and a relaxed gaze the entire cluster seems to

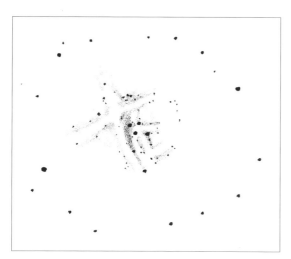

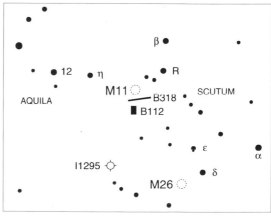

smolder with smoke. Try defocusing the telescope slightly and concentrating on the dark areas, and a most remarkable vision materializes: the shadow of a crucifix with light and dark rays radiating from its center. Look roughly one cluster diameter away from M26 in all directions and see if you don't pick out a "sacred circle" of similarly bright stars enclosing the cluster and cross.

M27

Dumbbell Nebula
NGC 6853
Type: Planetary Nebula
Con: Vulpecula
RA: 19^h 59^m.6
Dec: +22° 43′
Mag: 7.3
Dim: 8′.0×5′.7
Dist: 815 l.y.
Disc: Messier, 1764

MESSIER: [Observed 12 July 1764] Nebula without a star, discovered in Vulpecula, between the two forepaws, and very close to the fifth-magnitude star Flamsteed 14 in that constellation. It can be seen clearly in a simple three-and-a-half-foot refractor. It appears oval-shaped and does not contain any stars. M. Messier plotted the position on the chart of the comet of 1779, which will be published in the Academy volume for that year. Observed again 31 January 1781.

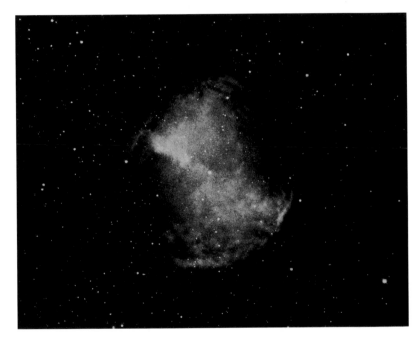

Although no star in Vulpecula shines brighter than magnitude 4.4, this summer constellation does boast the most famous of planetary nebulae, M27, the Dumbbell Nebula. Discovered by Messier in 1764, the Dumbbell got its nickname much later from its resemblance to a bodybuilder's hand weight. M27, a "splendid enigma," as Smyth described it in the *Cycle of Celestial Objects,* is one of the closer planetaries (815 light years away), and its physical diameter of 1.2 light years also makes it one of the larger. The gaseous material was blown from the blue-dwarf star now at its center during one of the star's death throes 48,000 years ago (which makes M27 more than twice as old as typical planetaries). M27 is a multiple-shell planetary. One shell, of doubly ionized oxygen, is expanding at a velocity

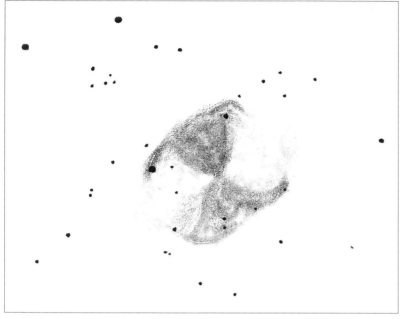

NGC: Magnificent object, very bright and large, binocular, irregularly extended (Dumbbell).

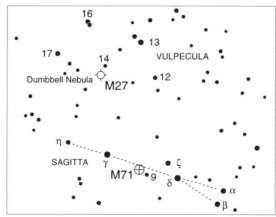

of 9 miles per second, while another shell, of ionized nitrogen, is expanding at 20 miles per second. From our vantage point, the whole gaseous ring is swelling 6″ per century. The gas shells glow from excitation by ultraviolet radiation emitted by the hot central star.

M27 is easily spotted in 7×50 binoculars as a roughly 7th-magnitude, 8′×6′ glow about 3° due north of 3.5-magnitude Gamma (γ) Sagitta, or less than ½° south of 5.7-magnitude 14 Vulpeculae. It is one of the few nebulous Messier objects whose visual impression truly matches that of its photographic image. At 23×, the brightest portion of the nebula shines with a pale green light and has a distinct hourglass shape (oriented northeast–southwest). Faint loops of nebulosity can also be seen extending to the northwest and southeast. In the 4-inch, these faint extensions vanish with increased magnification, because the surface brightness becomes too low for them to stand out against the sky background.

When seen in a 3° field of view, the Dubmbell Nebula appears to be standing upright on a long blanket of stars. Look carefully where the Dumbbell stands, and see if the star blanket does not seem to be sagging under the Dumbbell's weight. This wonderful illusion is created by clouds of dark nebulosity, which frame the star blanket. You will need a dark sky, though, to see this well. Now return to M27. The dumbbell-shaped gas cloud is part of a more complex structure – namely, the dumbbell sits inside a football-shaped shell of gas oriented 90° from it. There's a definite asymmetry: the southwestern lobe of the dumbbell is brighter than the northwestern lobe, while the northwestern side of the "football" is brighter than the southeastern side.

High magnification should reveal the planetary's turbulent interior. Look for several meandering swirls that create a marblelike texture among a field of foreground stars. Stare at the southwest lobe of the Dumbbell and you may see the clumps of light matter and dark matter coming together to form a wedge-shaped face – devilish almost, with curved, protruding horns.

But M27's central star is the real demon. Although stars of 14th magnitude are usually a cinch from pristine skies, glimpsing this one requires keen averted vision. The reason might lie in the difficulty of seeing stars through nebulosity. Using high magnifications usually solves this problem, but it doesn't seem to in this case. It is interesting that drawings of M27 made by Rosse, John Herschel, Smyth, and Leopold Trouvelot, who all used sizable telescopes a century or more ago, do not include the central star! Furthermore, a 1908 Lick Observatory catalogue logs the central star as magnitude 12, and Mallas described it as 13th magnitude. But, more recently, *Sky Catalogue 2000.0* lists the star's magnitude at 13.9, and Luginbuhl and Skiff, 13.8. It could be that the earlier magnitude estimates were just off a little, but has anyone investigated the possible variable nature of this star (or of its "17th-magnitude" companion)?

M28

NGC 6626
Type: Globular Cluster
Con: Sagittarius
RA: $18^h 24^m.5$
Dec: $-24° 52'$
Mag: 6.9
Dia: 10′
Dist: 20,000 l.y.
Disc: Messier, 1764

MESSIER: [Observed 27 July 1764] Nebula discovered in the upper part of the bow of Sagittarius, about one degree from the star λ, and not far from the beautiful nebula [M22] that lies between the head and bow. It does not contain any stars; it is circular and visible only with difficulty in a simple three-and-a-half-foot refractor. Its position was determined relative to λ Sagittarii. Observed again 20 March 1781.

NGC: Remarkable globular, very bright, large, round, increasingly compressed in the middle, well resolved, stars from 14th to 16th magnitude.

Like M23, M28 is another lost gem in the glittering Sagittarius Milky Way. This tiny (10′) but charming globular cluster suffers the misfortune of being too close to the bigger and overpoweringly beautiful globular M22. M28 lies only 1° northwest of 2.8-magnitude Lambda (λ) Sagittarii, the golden K2 star marking the top of the Sagittarius teapot, and can be seen in binoculars shining at 7th magnitude. Messier couldn't resolve any stars in the object (calling it a "nebula") upon discovering it in 1764 or when he reexamined it in 1781. His contemporary, William Herschel (who used much larger telescopes than Messier), could, and the object was correctly identified as a globular cluster in Smyth's 1844 *Cycle of Celestial Objects*.

At moderate power (72×), the cluster appears to have a tightly packed center surrounded by a diffuse halo of stars. The core shows a faint straw tinge, while the halo is pale blue. A "prominent," roughly 12th-magnitude star abuts the halo to the south, but the cluster is difficult to resolve in the 4-inch refractor. Still, I can start to resolve some of the brighter members (ranging from 13th to 14th magnitude) populating the halo, which seems significantly elongated to the west. In fact, with averted

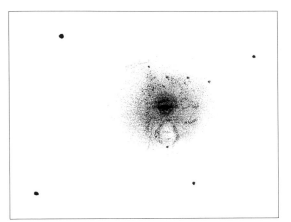

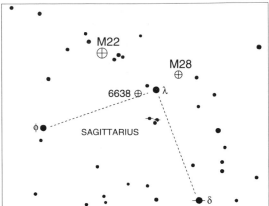

vision, the distribution of stars in the halo makes the cluster look boxy. Several times I glimpsed a dark "spike" piercing the halo from the southwest. To the north, brushes of starlight flare out from the halo, a characteristic I've seen with many other globulars.

The view is very different at 130×! A tiny, peach-colored nucleus is surrounded by several patches of stars. (Most globulars do not show color at high power, their color is more prominent at low power.) To me the core seems aged – those patches all look very old and weak. With averted vision, the halo teems with innumerable 13th- to 14th-magnitude stars. Through the Lick 36-inch refractor, I found the nucleus to be highly fragmented, with rafts of starlight seeming to drift away from an agitated core.

About ½° southeast of Lambda, you'll find the 9th-magnitude globular cluster NGC 6638. At 23×, NGC 6638, M28, and M22 can all just barely fit into the same field of view. If your telescope allows it, take some time to look at them together. Here are three globulars of vastly different visual proportions. M22 seems monstrous in comparison to minuscule NGC 6638, which spans only 5'; M22 is five times larger in apparent size! And do you see a color difference between M28 and M22?

M29

NGC 6913
Type: Open Cluster
Con: Cygnus
RA: 20^h 24^m.1
Dec: + 38° 29'.6
Mag: 6.6
Dia: 10'
Dist: 4,400 l.y.
Disc: Messier, 1764

MESSIER: [Observed 29 July 1764] Cluster of seven or eight very faint stars, which are below γ Cygni, and which look like a nebula in a simple three-and-a-half-foot refractor. Its position was determined from γ Cygni. This cluster is plotted on the chart for the comet of 1779.

NGC: Cluster, poor and little compressed, bright and faint stars.

M29 is a small packet of 80 stars contained in a tiny 10' sphere of sky. Still, this open cluster is visible to the naked eye as a 6.6-magnitude "star" just 2° south-southeast of 2.2-magnitude Gamma (γ) Cygni. It lies in a dense region of Milky Way that runs the length of the celestial Swan (Cygnus) – a region laced with dark lanes of interstellar dust. For small-telescope users, the cluster displays about a dozen 8th- and 9th-magnitude stars. It would no doubt be a more striking sight were it not for the dense obscuring matter that veils the region, dimming the cluster's light by three magnitudes!

I almost agreed with John Mallas, who said that each increase in magnification reduces the cluster's beauty. But when I used 130× and let my imagination fly, I saw a stream of 13th-magnitude stars coursing through the banks of brighter stars. Perhaps in larger instruments this stream would not have been so pretty, but it was quite alluring in the 4-inch, because it was so faint. Then something interesting happened. As my eye followed that elegant river, my mind drifted to a time when I stood waist high in sawgrass in the Florida Everglades. There my surroundings looked rather bleak, until I caught sight of a single orange wildflower. That tiny splash of color transformed a dull landscape into a grand sensation. Likewise, the delicate stream of stars in M29 turns an otherwise unremarkable cluster into a memorable one.

After plotting these stars, I returned to the eyepiece with the Everglades fresh on my mind. Now my eye played connect the dots, and I saw an entire scene – an aerial view of the glades. I noticed a bright wedge of Milky Way west of M29 and envisioned a flock of flamingos; and two stars to the east of M29 were crocodiles in the salty Atlantic. All this is

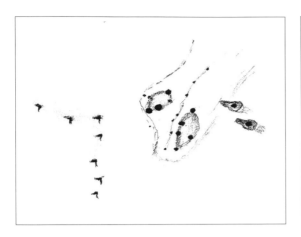

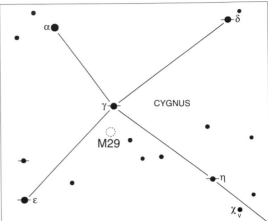

highly imaginative, but such imagery helps me to unite the wonders I see in the sky with those I've experienced on earth. Such fancy is the very foundation of celestial mythology. Why not create your own mythology with the star patterns in open clusters?

M30

NGC 7099
Type: Globular Cluster
Con: Sagittarius
RA: 21ʰ 40ᵐ.4
Dec: –23° 10′
Mag: 6.9
Dia: 12′
Dist: 26,700 l.y.
Disc: Messier, 1764

The thirtieth entry in Messier's catalogue is easily spotted with binoculars about ½° west-northwest of 41 Capricorni – a 5.5-magnitude star 6½° east-southeast of 3.7-magnitude Zeta (ζ) Capricorni. M30 is a fairly large globular cluster, measuring 93 light years in diameter, that whizzes through

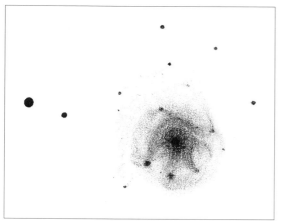

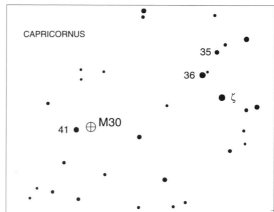

space at 108 miles per second in approach. Visually compressed, M30 has a tiny 1′ core inside a 12′ globular haze. Despite the object's brightness (magnitude 6.9), low power does not resolve it at all.

At 72× the view is most peculiar, as if whoever had started to mold the cluster's shape suddenly decided to stop, never finishing. It has a lopsided appearance; the northern portion displays well-defined fingers of stars, while the southern part looks kneaded but incomplete. Smyth saw this too: "From the straggling streams of stars on its northern verge, [it] has an elliptical aspect, with a central blaze."

At high power the entire cluster seems to be served on a plate of fainter stars. The northern jets of stars contain bright stellar knots, all between 12th and 13th magnitude, which seem to flow downwind of a gale. More than anything, I'm reminded of the nucleus of a very active comet. (Webb, too, thought M30 was cometlike.) Be sure to spend time concentrating on the core, which is rather corkscrew shaped. In his *Astronomical Objects for Southern Telescopes*, E. J. Hartung writes that M30's "well-resolved centre is compressed and two short straight rays of stars emerge [north preceding] while from the N edge irregular streams of stars come out almost spirally."

Return to low power. Now doesn't the cluster look as if it has horns to the north? The challenge is to see a very faint extension of stars on the southeast halo and its very faint semicircular wings.

MESSIER: [Observed 3 August 1764] Nebula discovered below the tail of Capricornus, close to the sixth-magnitude star Flamsteed 41. It is difficult to see with a simple three-and-a half-foot refractor. It is circular and does not contain any stars. Its position was determined relative to ζ Capricorni. M. Messier plotted it on the chart for the comet of 1759, *Mémoires de l'Académie 1760*, plate II.

NGC: Remarkable globular, bright, large, slightly oval. From its edge, it gradually brightens to a much more intense middle. Stars from 12th to 16th magnitude.

M31

Andromeda Galaxy
NGC 224
Type: Spiral Galaxy
Con: Andromeda
RA: $0^h 42^m.7$
Dec: $+41° 16'$
Mag: 3.4
SB: 13.6
Dia: 3°x1°
Dist: 2.3 million l.y.
Disc: Persian astronomer Al-Sufi, tenth century

MESSIER: [Observed 3 August 1764] The beautiful nebula in the belt of Andromeda, shaped like a spindle. M. Messier examined it with several instruments, but he was not able to detect any stars. It resembles two cones or pyramids of light, joined at their bases, and the axis of which lies northwest to southeast. The two points of light are perhaps some 40 minutes of arc apart. The common base of the two pyramids is about 15 minutes. This nebula was discovered in 1612 by Simon Marius, and has been observed subsequently by various astronomers. M. le Gentil gives a drawing of it in *Mémoires de l'Académie 1759*, page 453. It is plotted in the English *Atlas Céleste*.

NGC: Magnificent object, extremely bright, extremely large, very much extended.

To the true romantic of astronomy, M31 will always be known as the "Great *Nebula* in Andromeda" – a name bestowed upon it before spectroscopy revealed that this luminous mist was not the protoplasmic soup of a solar system in formation but a distant island universe like our own Milky Way Galaxy. An enormous pinwheel of dust and gas, the Andromeda Galaxy contains some 300 billion suns spread across 130,000 light years. It is rushing toward us at 185 miles per second. M31 is among the largest galaxies known and is by far the largest member of the Local Group of galaxies, which includes our Milky Way and some two dozen smaller systems. The Andromeda and Milky Way galaxies dominate the Local Group with their size, with M31 being twice as massive as our Milky Way. And though we see the Andromeda Galaxy nearly edge on, astronomers see enough structure to speculate that the Milky Way is similar in shape and structure. If you were in the Andromeda Galaxy looking at the Milky Way, the Milky Way would appear much the same way as M31 does to us.

At 2.3 million light years distant, M31 is also one of the farthest objects visible to the naked eye. Under reasonably dark skies it appears as

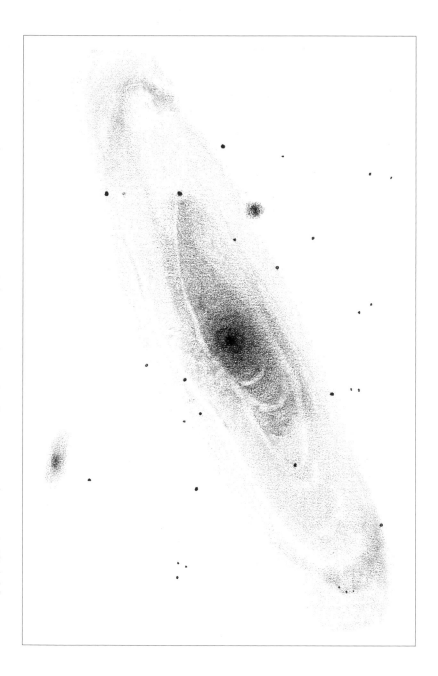

a cocoon of nebulous vapor 1° west of 4.5-magnitude Nu (ν) Andromedae in the Chained Maiden's belt. M31 stretches 3°, or nearly 6 moon diameters, on the most transparent nights. A good pair of binoculars will show some of the galaxy's subtle detail. Even 7×35 binoculars will reveal its elliptical disk, whose surface brightness gradually fades away from a star-

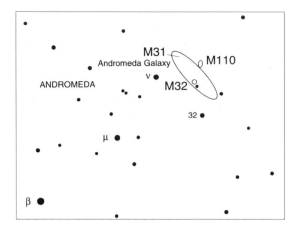

like core. The billions of suns in the core are so tightly packed that astronomers believe there may be a black hole at the center.

The galaxy's northwestern rim has a sharp edge to it, which marks the location of a prominent dark lane slicing through that part of the galaxy; in contrast, the galaxy's southeastern rim diffuses gradually into the sky background. Binoculars will also reveal two of M31's companion galaxies: M32 and M110. M32 looks like a slightly swollen 8th-magnitude star on M31's outermost bright rim, roughly ½° south and slightly east of the nucleus. M110 is a similarly bright, though larger, elliptical haze 37′ northwest of M31's nucleus. For more on M110, see pages 282–3.

In the 4-inch at 23×, a bright yellow "star" marks the very center of the Andromeda Galaxy. It lies inside several tightly wrapped pale yellow haloes, which start out circular close to the core but become progressively more elliptical and more skewed toward the southwest farther away from the nucleus. This teardrop-shaped patch of golden light is surrounded by an enormous ashen elliptical halo grooved with faint dust lanes. The detailed drawing on the preceding page is a composite: after spending three hours each night over several nights examining the galaxy, I combined the views made with low, medium, and high powers. I concentrated on the nuclear region the first night, the northeast portion on the next night, and the southwest sector last; the two satellite galaxies were viewed on separate nights.

Besides the prominent nuclear region, the most striking feature of M31 is a conspicuous dust lane running along its northwest edge. One night I followed that lane southwestward for half the galaxy's radius before it looped back in classic spiral fashion. When I used 130× to study the portion of this lane lying closest to the nucleus, it appeared very turbulent. One could spend hours trying to visualize patterns in the lacy swirls and splotches of dust. See if you can detect a leopard-spot pattern there; it

M32

NGC 221
Type: Dwarf Elliptical Galaxy
(companion to M31)
Con: Andromeda
RA: $0^h 42^m.7$
Dec: $+40° 52'$
Mag: 8.2
SB: 12.7
Dim: $8'.7 \times 6'.5$
Dist: 2.3 million l.y.
Disc: Guillaume Le Gentil, 1749

MESSIER: [Observed 3 August 1764] Small nebula without stars, below and a few minutes away from the nebula in the belt of Andromeda. This small nebula is circular, its light fainter than that in the belt. M. le Gentil discovered it on 29 October 1749. M. Messier saw it for the first time in 1757, and has not noted any change in its appearance.

NGC: Remarkable, very bright, large, round, suddenly much brighter in the middle toward the nucleus.

almost blends with another dust lane farther to the northwest. Now look just east of the innermost halo surrounding the nucleus. There you should find another lane of dust with two faint dark hooks branching off it to the northeast. If you're having trouble with high magnification, revert to low power, because you will condense the galaxy's light and increase the contrast between the bright haze of the galaxy and the silhouettes of dust.

The arms in M31's outer halo contain some bright concentrations, which, with imagination, look like spits of gray sand between streams of dark matter. Jones said these show best in photographs sensitive to blue light. Interestingly, my eyes appear to be sensitive to blue light, so I can see these concentrations well. English nova and comet discoverer George Alcock is also believed to have blue-sensitive eyes. Other observing friends, like Michael Mattei of Harvard, Massachusetts, have red-sensitive eyes. One way you can test your eyes for color sensitivity is to observe astronomical objects with blue or red features and see how you fare. You can start with these blue concentrations in M31's spiral arms. Another good one is NGC 206 close to the galaxy's southwestern rim. This $1' \times 2'$ patch of fuzzy light is actually an enormous star cloud within the galaxy that measures $700 \times 1,300$ light years. Based on my limiting-magnitude studies, about a dozen globular clusters are within range of a good 4-inch telescope under perfect conditions. By the way, have you ever noticed how the extreme tips of M31's outer spiral arms curve away from the main body? The southwestern tip curves to the south, while the northeastern tip curves north.

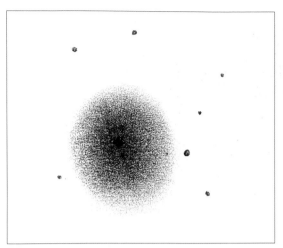

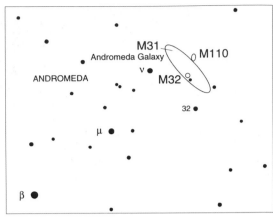

M32 is easy to mistake for a bright star in binoculars. At low and moderate powers in the telescope, M32 is essentially a featureless circular glow. But take the time to look for a faint outer envelope that gives this seemingly round galaxy its elliptical shape. At $130\times$, I could discern a definite starlike core with an odd vertical extension running northeast to southwest through the entire galaxy. Is this a real feature or an illusion created by a peculiar alignment of faint, unresolved foreground stars? The southeastern side of the nucleus seems to be bordered by an arc of bright haze, while the same distance away to the northeast, there appears to be a faint star that flickers in and out of view. I like to see it as a beacon, a friendly message from a distant neighbor.

Like M31, M32 is believed to harbor a black hole, but one perhaps 10 times smaller than the one at the heart of M31. Also like M31, M32's light is blue-shifted, but this system appears to be approaching us at a slower speed of 126 miles per second.

M33

Triangulum Galaxy or
Pinwheel Galaxy
Type: Spiral Galaxy
NGC 598
Con: Triangulum
RA: $1^h 33^m.9$
Dec: $+30° 39'$
Mag: 5.7
SB: 14.2
Dim: $71' \times 42'$
Dist: 2.3 million l.y.
Disc: Messier, 1764

MESSIER: [Observed 25 August 1764] Nebula discovered between the head of the northern Fish [in Pisces] and Triangulum, close to a sixth-magnitude star. The nebula's light is whitish, and almost even in density, but is slightly brighter over the central two-thirds of its diameter, and it does not contain any stars. It is difficult to see with a simple one-foot refractor. Its position was determined relative to α Trianguli. Observed again 27 September 1780.

NGC: Remarkable, extremely bright and large, round, very much brighter in the middle to a nucleus.

In long-exposure photographs M33, another member of our Local Group of galaxies, looks like an enormous spiral with innumerable suns clinging to wildly spinning arms. But with a diameter of 50,000 light years, you could easily fit three M33s in the disk of M31. In fact, M33 may be a satellite galaxy of M31, orbiting it just as the moon does the earth. The Andromeda Galaxy is also about 15 times more massive than M33, which is about two times smaller and seven times less massive than our Milky Way. As seen from an imaginary planet in the Pinwheel Galaxy, M31 would

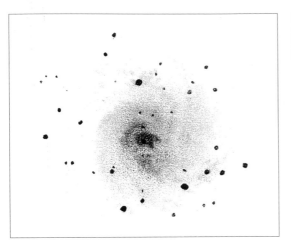

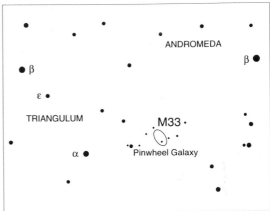

be an impressive sight – an oblique swarm of faintly glittering suns stretching 6° in that hypothetical sky.

Regardless of its true size, the Pinwheel is a great sight from earth. It has long been a naked-eye challenge for amateur astronomers. While some find it easily visible to the naked eye or in binoculars, others cannot see it at all. The problem lies in the galaxy's low surface brightness. Although the total magnitude of M33 is the same as a 6th-magnitude star, the galaxy's light is spread over an area of sky larger than two full moon diameters, making it appear dim. A dark sky, a steady atmosphere, and good vision are required to see M33 with or without optical aid. Its ease of visibility is, as Walter Scott Houston often stated, a barometer for the clarity of one's observing site. The galaxy is completely washed out in urban skies and can be disappointingly dim from suburban locations even through large-aperture telescopes! Try to see it, though, without optical aid, because M33 is one of the farthest objects visible to the naked eye. Some observers claim M33 cannot be seen with the naked eye at all and that the "glow" is caused by some nearby stars. But the brightest naked-eye stars near M33 are 7th-magnitude and fainter, so that argument is specious.

The clockwise-swirling spiral of M33 can be spotted 4° west-north-west of 3.4-magnitude Alpha (α) Trianguli. M33 and M31 lie on opposite sides of, and at nearly equal angular distances from, the 2nd-magnitude red giant Beta (β) Andromedae. Binoculars do a fine job of bringing out the Pinwheel's luster against the black backdrop. Even 7×35s show the 50,000-light-year-wide galaxy as a distinct oval glow immediately north of an 8th-magnitude star. At 23× in the 4-inch, the galaxy's light is compressed into a shimmering disk of optimal contrast, with several faint spiral arms sweeping away from a tight, lens-shaped nuclear region. Take a moment to compare the size of that tiny nucleus to the rest of the galaxy; M33's nucleus contains less than 2 percent of the galaxy's total mass. In comparison, M31's nuclear region is about one twenty-fifth the size of the entire galaxy and about five times larger than M32!

Although John Herschel called M33 unfit for high powers, "being imperceptible from want of contrast with 144×," I find high magnification (130×) perfect for concentrating on its tiny central knot, which appears misshapen by dark matter. Several faint stars or fuzzy kinks lurk in the misty vicinity of the galaxy's core. Using a 12-inch telescope at 225×, Luginbuhl counted perhaps as many as eight 13th-magnitude stars sprinkled in the southern portion of this region. "The whole surface of the galaxy," he writes, "is covered with faint stellarings and splotches." Similarly, Rosse saw it full of knots with two S-shaped curves crossing in the center, though only one-long and loose S is conspicuous at low power. But, again, high power will reveal some taut and stubby inner spiral structure.

Without question, M33 is the easiest galaxy beyond the Magellanic Clouds to resolve with large amateur instruments. Skiff notes that many faint clusters, stellar associations, and nebulae embedded in it are within grasp of a 10-inch telescope. I recorded in my observing diary that M33's most prominent arms appear lumpy at 23×, and with 72× the contrast holds up well enough for there to be a suspicion of resolution. M33's largest HII region (NGC 604) is by far the most conspicuous feature in the galaxy besides the nucleus. You'll find it 12′ northeast of the nucleus, just 1′ northwest of a 10.5-magnitude star. Luginbuhl saw this feature in a 2.4-inch (60-mm) telescope as a concentrated spot in the halo! Can you see it in binoculars?

M34

NGC 1039
Type: Open Cluster
Con: Perseus
RA: 2^h 42^m.1
Dec: + 42° 45′
Mag: 5.2
Dia: 25′
Dist: 1,450 l.y.
Disc: Messier, 1764

MESSIER: [Observed 25 August 1764] Cluster of faint stars, between the head of Medusa [in Perseus] and the left foot of Andromeda, slightly below the parallel of γ. The stars may be detected with a simple three-foot refractor. Its position was determined from β [Persei] in the head of Medusa.

NGC: Bright, very large cluster, little compressed, of scattered 9th-magnitude stars.

Cast your sights on the constellation Perseus and what first catches your eye is the large misty splotch of the Double Cluster (NGC 869 and NGC 884), which in binoculars resolves into a grand pair of bright stellar splatterings a half-degree apart. While it is curious that the Double Cluster was overlooked by Messier for his catalogue (was it too obviously not a comet?), he did discover another open cluster, now called M34, glimmering nearby, just north-northeast of the midpoint between the famous variable star Algol (Beta [β] Persei) and the beautiful double star Gamma[1,2] ($γ^{1,2}$) Andromedae. M34 is a loose aggregation of stars about 100 million years old (much older than either of the Double Cluster components). Its 60 members are spread across a scant 10 light years of space. Regardless, Webb called it one of the finest objects of its class.

Shining at magnitude 5.7, M34 is clearly visible to the naked eye. Note how much easier M34 is to see than M33. Although they have the same magnitude, M33 is about three times larger in apparent diameter, so it has a lower surface brightness. M34's scattering of bright pearls includes about a dozen suns that shine brighter than 9th magnitude, all of which can be resolved with 7×35 binoculars; several of these are white giants. What looks like the brightest star in M34 shines at magnitude 7.3, but this star is a foreground star, not a true cluster member. One of the brightest true members is a double star known as Struve 44, whose 8.4- and 9.1-magnitude components are separated by 1.4″.

Although M34 is certainly a first-rate cluster for small telescopes, Walter Scott Houston preferred the view through 15×65 binoculars. More magnification, he said, merely spreads out the few bright stars that the binoculars show perfectly well.

Still, the view of the cluster's core at $72 \times$ is quite dramatic. A spray of double stars, many of similar brightness, appear to be fleeing from two 8th-magnitude Type $A0$ gems separated by 20″ (the double star h1123) at the very heart of the cluster. Smyth also noted the gathering of "coarse pairs" in this "scattered but elegant group." In fact, he seemed to favor the central double more than the cluster itself, calling M34 a "double star in a cluster."

After probing the depths of the cluster, return to low power, then slightly defocus the view and look for two large stellar arcs abutting the cluster's core to the southwest (see the drawing). If you visually measure the radius from the core to the outer arc, then look about the same distance to the northeast: you may see some hazy outer arcs of starlight. They might show more clearly if you gently sweep the telescope back and forth over this region. Once you see the arcs, use averted vision to stare at them, and they will resolve into individual stars. The region between the faint arcs and the core looks rather vacuous. But this blank area is actually filled with very faint background stars.

Algol, the bright star about 5° to the southeast of M34, is one of the most famous variable stars in the night sky. Its placement in the heavens represents the head of Medusa, the serpent-haired Gorgon of classical mythology, held by her slayer Perseus. The "demon star" usually shines at magnitude 2.1, but nearly every three days it mysteriously fades to magni-

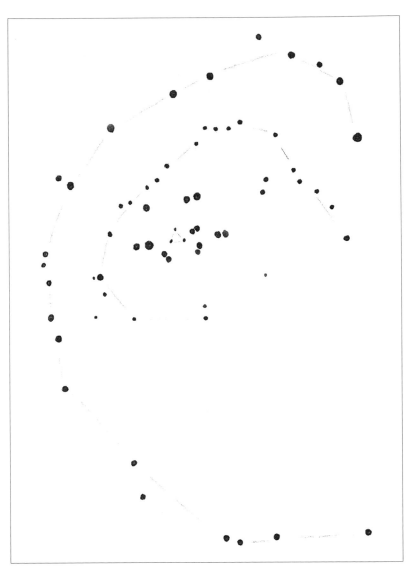

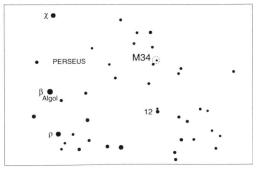

tude 3.4, before brightening again, all in just 10 hours. This periodic dimming occurs because Algol is an eclipsing binary star – a pair of stars orbiting a common center of mass – whose orbital plane lies in our line of sight. Every 2 days 20 hours 48 minutes and 56 seconds, Algol's larger but comparatively dim secondary star eclipses its smaller but brighter primary by 79 percent. We see this drama unfold as an apparent winking of the star. By the way, the dimming is not constant, in part because Algol is not a simple binary star, but a complicated quadruple system.

NGC 891, a more challenging but worthwhile object in Andromeda, lies 3° to the west-southwest of M34, about midway between it and Gamma Andromedae. This 10th-magnitude spiral galaxy ($13' \times 2'.8$) is seen exactly edge on. Furthermore, the system has a diameter of about 160,000 light years, about 1½ times larger than our Milky Way Galaxy. But we can still imagine NGC 891 as our galaxy seen edge on about 43 million light years distant.

M35

NGC 2168
Type: Open Cluster
Con: Gemini
RA: 6^h 09^m
Dec: + 24° 21'
Mag: 5.1
Dia: 25'
Dist: 2,800 l.y.
Disc: uncertain; probably Philippe Loys de Chéseaux, 1745

"A marvelously striking object: no one can see it for the first time without exclamation." William Lassell, a nineteenth-century English amateur astronomer, penned this description of M35 in Gemini based on a view through his 24-inch reflector. But this bright 5th-magnitude open cluster is equally exquisite when seen through small apertures. To the naked eye

M35

MESSIER: [Observed 30 August 1764] Cluster of very faint stars, close to the left foot of Castor [the western twin of Gemini], not far from the stars μ and η in that constellation. M. Messier plotted its position on the chart for the comet of 1770, *Mémoires de l'Académie 1771*, plate VII. Plotted in the English *Atlas Céleste*.

NGC: Cluster, very large, considerably rich, pretty compressed, stars from 9th to 16th magnitude.

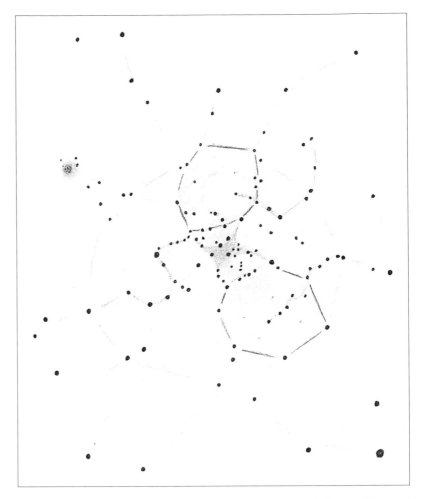

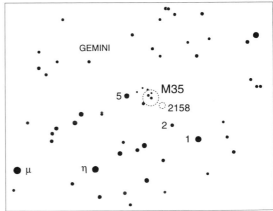

Deep-Sky Companions: The Messier Objects

it is a mottled splash of hazy light with an apparent diameter nearly that of the full moon (the cluster's true diameter is about 20 light years). It has a decidedly rectangular shape with a bright eastern ridge and a seemingly tight core. The mottled naked-eye appearance is almost aggravating, because no amount of time seems enough to resolve it faithfully. The cluster's brightest members shine between 8th and 9th magnitude and, not surprisingly, lie along that conspicuous eastern wall (actually an arc). Someone with keen, young eyes might be able to resolve a few of the stars in the wall from a good site at high elevation. (Although I can convince myself that the cluster is resolved, I cannot pinpoint the location of any given star accurately enough to prove it; and therein lies the challenge.)

The outlying stars fill a 1.5° field and extend east to the 6th-magnitude star 5 Geminorum – a view best appreciated through a rich-field telescope. When I visually connect the brightest stars in this extreme outer halo, they resemble a scallop shell's wavy edges. The sides of the shell are marked by 5 Geminorum to the east-northeast and the tiny 8.6-magnitude open cluster NGC 2158 to the southwest. A snaking dark river flows between the scallop's halves, making them appear slightly open.

Star counts in M35 vary dramatically with the observer. Rosse counted 300 in a field of 26'. Luginbuhl and Skiff tally 200-odd stars in a field of 30'. Åke Wallenquist counted 119 stars in a diameter of 30', estimating the stellar density there to be about 6 stars per cubic parsec, which is very loose. (By comparison, the density of M34's core is 21 stars per cubic parsec.) The central stars form a figure eight pattern. A corset of about two dozen stars seems to hold in the cluster's slender waist. This central girdle is flanked by two large voids: one to the south, the other to the northeast. Is this absence of stars caused by obscuring dust in the galactic arm? Look carefully at these voids and see if there isn't a wavy stream of stars originating at the center of each of them. Both streams flow to the east, though the southern one makes a sudden curve to the south. If you concentrate, you might see these streams coming toward you, an illusion that adds dimension to the otherwise flat cluster.

Star streams also radiate in all directions from the figure eight. They connect to an outer ring of similarly bright stars. This view mimics the smoky remains of a fireworks display faintly illuminated by city lights. The nineteenth-century observer Smyth had a similar impression of M35, saying its pattern of stars reminded him of a bursting skyrocket.

Earlier I mentioned NGC 2158; it's a pint-size star cluster only 5' in diameter that lies ½° southwest of M35. Although NGC 2158 appears smaller and dimmer than M35, the two clusters are really very similar in physical size; it's just that NGC 2158 is six times more distant. Visually, NGC 2158 bears some resemblance to NGC 1907 near M38 (see page 129);

both clusters appear as milky spots with some stars hovering near the limit of vision.

Burnham seems to have severely underestimated the brightness of both NGC 2158 and the brightest stars within it. He estimated the cluster to be 11th magnitude with the brightest members being magnitude 16. I was surprised to find the cluster's milky haze quite easily in the 4-inch, and at 130× I resolved some of its brighter members. Skiff determined the cluster's visual magnitude to be 8.6, and through an 8-inch telescope he resolved some 50 stars of magnitude 13 or fainter. NGC 2158 is very sensitive to light pollution. I never could see this object through Harvard's 9-inch refractor because of urban skyglow.

M36

NGC 1960
Type: Open Cluster
Con: Auriga
RA: $5^h 36^m.3$
Dec: $+34° 08'.4$
Mag: 6.0
Dia: 10'
Dist: 4,100 l.y.
Disc: Guillaume Le Gentil, 1749; Giovanni Batista Hodierna noted it before 1654

MESSIER: [Observed 2 September 1764] Star cluster in Auriga, close to the star φ. With a simple three-and-a-half-foot refractor it is difficult to distinguish the stars. The cluster does not contain any nebulosity. Its position was determined from φ.

NGC: Cluster, bright, very large and rich, little compressed, with scattered 9th- to 11th-magnitude stars.

M36 is one of three bright Messier clusters decorating a rich band of the Auriga Milky Way (the others being M37 and M38). To find it, look about 8° east of 2.7-magnitude Iota (ι) Aurigae, or about 6° north-northeast of 1.6-magnitude Gamma (γ) Aurigae. M36 and its closest neighbor, M38, appear as hazy "stars," pieces of lint clinging to the dusty veil of the Milky Way. A moderately young 20- to 30-million-year-old cluster, M36 has a true diameter of about 12 light years, with perhaps 60 stars visible in a field of 10'.

Telescopically, M36 appears much as Webb characterized it: a "beau-

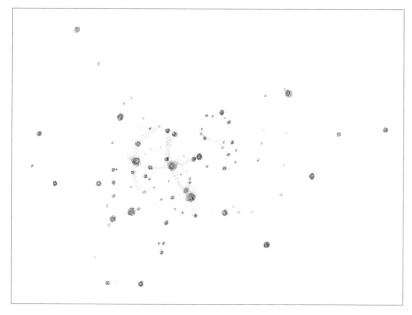

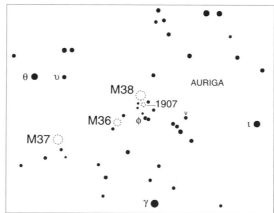

tiful assemblage of stars . . . very regularly arranged." Its 15 brightest members (between 9th and 11th magnitude) shimmer against a tight, hazy background caused by the feeble light of dozens of 13th- and 14th-magnitude stars. As with other open clusters embedded in the Milky Way's, it's hard to discern the actual outer boundaries of the cluster.

At 23×, M36 displays a loose central concentration of stars with two long, slightly curved arms (extending southwest and northeast) and smaller appendages perpendicular to them. D'Arrest saw the central swirls arranged in three tight spirals. Mallas envisioned a crab, and Jones imagined a rocking chair or a miniature Perseus. I see the central association forming a warped cross – one with a strikingly similar orientation to Webb's Cross at the heart of M38 (see page 129). Since the two clusters are neighbors, hop from one to the other and see if you can't make out the twin crosses.

Switch to medium power to examine the pair of close double stars straddling the cluster's center. The southerly one is Struve 737, a pair of 9th-magnitude stars separated by 11″. The wider, brighter double to the north is oriented about 90° from Struve 737, and (if you let your mind relax) it looks closer but really is not. It's just smoke and mirrors.

Now return to low power. At 23×, M36 shares the same wide-field view with M38. Place the two clusters so that they are on the northeast edge of a wide-field eyepiece, so you can see 5th-magnitude Phi (φ) Aurigae and its attendant stars forming a triangle with them 1½° to the west-southwest. Now sweep the telescope generously back and forth, moving first northeast to southwest, then northwest to southeast. Can you make out the lines of 8th- and 9th-magnitude stars connecting the clusters to one another and to Phi? These stars encompass a delta of Milky Way, which shines with the misty glow of countless faint suns, tiny clusters, and patches of nebulosity. It is one of the most rewarding views for rich-field telescopes in the northern skies. Large binoculars will also do justice to this region. Burnham found the cluster makes its best impression in 6- and 8-inch telescopes at 20× to 60×.

M37

NGC 2099
Type: Open Cluster
Con: Auriga
RA: 5ʰ 52ᵐ.3
Dec: +32° 33′.2
Mag: 5.6
Dia: 15′
Dist: 4,400 l.y.
Disc: Messier, 1764; Giovanni Batista Hodierna recorded it before 1654

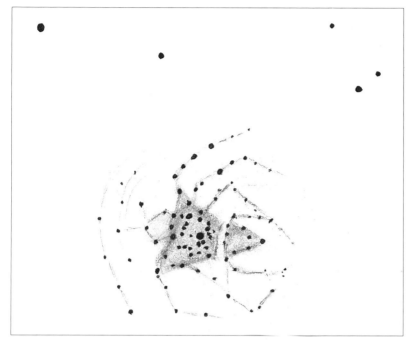

MESSIER: [Observed 2 September 1764] Cluster of faint stars, close to the previous one [M36], on the same parallel as χ Aurigae. The stars are fainter, closer together and are enveloped in nebulosity. It is difficult to see the stars with a simple three-and-a-half-foot refractor. This cluster is plotted on the chart of the second comet of 1771, *Mémoires de l'Académie 1777*.

NGC: Cluster, rich, pretty compressed in the middle, with large and small [bright and faint] stars.

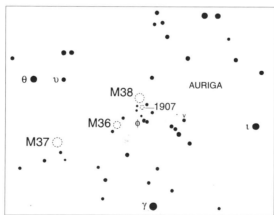

Locate bright Theta (θ) Aurigae, then use binoculars to follow a gently meandering stream of 7th- and 8th-magnitude stars flowing about 5° south-southwest of it. The stream pours into M37, a solitary pool of subtle beauty. Put down the binoculars and you should be able to pinpoint the open cluster's 5.6-magnitude glow with the naked eye. It is a small cluster (15′), so its light is compact. Telescopically, M37 is much more wonderful a sight than either M36 or M38. After looking at the two other clusters, M37's grandeur took me by surprise. At 23× this cluster looks like a finely resolved globular cluster – a marvelous illusion. Indeed, Luginbuhl and

Skiff likened it to a "broken down view" of M22 or Omega (ω) Centauri, and these are two of the greatest globulars in the night sky. The 200-million-year-old cluster contains some 1,890 members. About 500 of these suns are brighter than 15th magnitude, and they are spread across 15′ (about 20 light years of space).

Anyone who likes gemstones should appreciate M37, a 9th-magnitude topaz jewel surrounded by a pear-shaped cluster of scintillating diamonds. Smyth called it a "magnificent object; the whole field being strewn . . . with sparkling gold dust." The color of the central star alone could mesmerize even the most tenured observer, though the degree of color perceived varies with the observer. Some have referred to it as ruby red (another gem!), others as simply red, still others as pale red. Could this star be variable? Someone with a photometer should try monitoring the brightness and color of this intriguing star over the course of a year.

I am surprised that I haven't encountered more references in the literature to the obvious dark lane slicing through the eastern half of the cluster's core. Luginbuhl and Skiff described it as a dark void about 5′ across containing a single 12th-magnitude star. In this fashion, the pear-shaped diamond really looks more like a coat of arms. To enhance the dark lane's visual impact, slightly defocus the telescope. I have also seen this feature in photographs; try holding the photograph on page 00 at arm's length and squinting.

At moderate magnification M37 is gloriously detailed. Its stellar richness accounts for only half of the cluster's beauty, though. The other half is the vast network of dark lanes, which, at least from dark skies, becomes visually overpowering the longer I stare at the cluster. When I relax my gaze, the stars seem to fade into the background, while the dark lanes move forward. At that point, I feel as if I'm looking at the silhouettes of leafless trees against a sparkling expanse of Milky Way. If my eye jumps around the field, dark lines appear to slash across the cluster from all directions; it's as if I'm looking at this cluster through a severely fractured eyepiece. The illusion is even more dramatic than that of the dusty cobwebs covering M4.

If you consider the "legs" of stars extending north and south from M37's pear-shaped body, the cluster mimics the shape of a water strider insect in a puddle of rain, as my drawing depicts. I suspect these legs represent part of the "wonderful loops and curved lines of stars" seen by Rosse.

M38

NGC 1912
Type: Open Cluster
Con: Auriga
RA: 5^h 28^m.7
Dec: +35° 51'.3
Mag: 6.4
Dia: 15'
Dist: 4,200 l.y.
Disc: Guillaume Le Gentil, 1749; Giovanni Batista Hodierna noted it before 1654

MESSIER: [Observed 25 September 1764] Cluster of faint stars in Auriga, close to the star σ, not far from the two preceding clusters [M36 and M37]. This one is rectangular in shape and contains no nebulosity, if examined carefully with a good telescope. It may extend for about 15 minutes of arc.

NGC: Cluster, bright, very large and rich, with an irregular figure, large and small [bright and faint] stars.

About 2° northwest of M36 and faintly visible to the naked eye is the third Messier open cluster in Auriga, M38. At 15' in diameter, M38 is one-third larger than M36 in apparent size; it is also a half-magnitude fainter, independent of size. Surprisingly, M38 also has twice the number of stars that M36 has. M38 is one of the older galactic clusters and contains about 120 stars in an area equal to half the diameter of the full moon. The density of its loosely packed center is about 8 stars per cubic parsec – half the density of M36's core and only one-fifth that of M37's.

About a dozen of the brightest stars in M38 are arranged in a distinct cross at the center of the cluster. Webb saw this first as an oblique cross with a pair of brighter stars in each arm, a view that reminds me of a silver crucifix studded with jewels. This is most obvious at high power. Look carefully and you will see a ring of stars at the center of the cross within which shines a solitary 11th-magnitude star (dimmer stars can be seen but they are not obvious). A fainter ellipse of stars surrounds the cross, and, with imagination, I can see the arms of the cross wrapped around a hemispherical shell. This evokes a strong image of the bleached shell of a sea urchin seen at a slightly oblique angle, as shown in my drawing.

A ring of bright stars connects M38 to NGC 1907, a compressed 8th-magnitude open cluster 30' to the south-southwest. (Interestingly, this connection is not obvious at moderate to high powers or in small fields of view.) It doesn't take much imagination to at least contemplate M38 and NGC 1907 as being one cluster with a mighty void at the center (remember the black heart of M35?). I like to imagine I am looking at an aerial view of a galactic impact site, as if some hideously large comet once collided

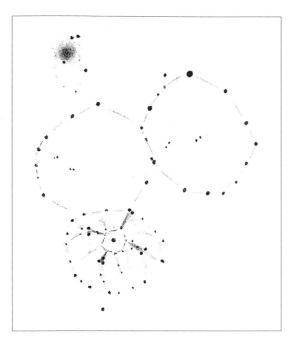

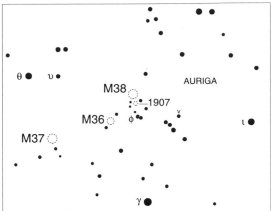

obliquely with the Milky Way in Auriga – the ejecta splashing out to the north-northeast to form M38, and a pile of rubble falling to the south-southwest to form NGC 1907. Move the telescope to the southeast where a second ring of stars adjoins the first. Inside each ring you'll find two pairs of tiny doubles (see drawing for locations).

Examine NGC 1907 at high power. Its dark veins make it appear like a trifid of stars. I really like this cluster because, through the 4-inch, at all powers, it looks like a tiny puff of smoke blowing out of a volcano. But this is an immediate impression. With careful study, it vaguely resolves into a rectangle of equally bright stars. It's rather strange to see such a uniform smattering of stars packed together in so small an area.

M39

NGC 7092
Type: Open Cluster
Con: Cygnus
RA: 21^h 32^m.2
Dec: +48° 26′.6
Mag: 4.6
Dia: 30′
Dist: 950 l.y.
Disc: Guillaume Le Gentil, 1750; Aristotle noted it around 325 B.C.

MESSIER: [Observed 24 October 1764] Cluster of stars near the tail of Cygnus. They can be seen with a simple three-and-a-half-foot refractor.

NGC: Very large, very poor cluster, very little compressed, of 7th- to 10th-magnitude stars.

Of the dozens of bright clusters and nebulae in Cygnus that are available to small telescopes, surprisingly few were recorded by Messier and his contemporaries – only three (M29, M39, and M56). And open cluster M39, near the Swan's tail, seems a most unlikely candidate for Messier's list because it is so large (30′) and sparse (only 30 stars).

Yet, shining at 4.6 magnitude, M39 does greet the naked eye as a

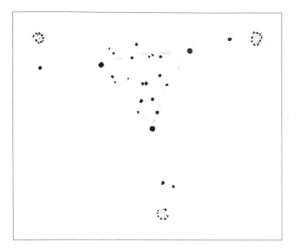

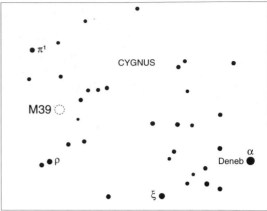

small fuzzy patch of light about 9° east-northeast of Deneb, or 3° north of 4th-magnitude Rho (ρ) Cygni. Even Le Gentil, who discovered the object in 1750, noticed that it could be seen without a telescope. At least three of its brightest members can be distinguished with the unaided eye, and 7 × 35 binoculars easily resolve about a dozen stars in an area of sky the size of the full Moon. Smyth seemed pleased with this venerable 300-million-year-old cluster, calling it a "splashy field of stars in a rich vicinity."

The beauty of M39 lies in the uniform brightness of its most prominent members, their diamond blue purity, and the cluster's geometry – a triangle of starlight with a fine double star at its heart. Smyth also noted, as the drawing shows, that many of the stars are in pairs. I do wonder about Flammarion's description: "unusual curved runners of stars, with a compressed cluster of 20 stars, difficult to separate from the rest." My guess is that he based his comment on a binocular view, because I too have noticed one significant curve of stars running through that triangle in binoculars.

Meanwhile, use low power to survey the area around M39. Do you notice anything curious? Look outward from each of the triangle's three corners and you will find a tiny asterism of stars. Together the three asterisms form a perfect triangular encasement for the smaller triangle of M39.

M40

Winnecke 4
Type: Double Star
Con: Ursa Major
RA: $12^h 22^m.2$
Dec: $+58° 05'$
Mag: 9.0 and 9.6
Sep: 49″
Dist: −
Disc: John Hevelius, 1660

MESSIER: [Observed 24 October 1764] Two stars very close to one another and very faint, located at the root of the tail of Ursa Major. They are difficult to detect in a simple six-foot refractor. It was while searching for the nebula that lies above the back of Ursa Major as plotted in the book *Figure des Astres*, and whose right ascension should have been 183° 32′ 41″ and declination +60° 20′ 33″ in 1660 – which M. Messier could not see – that he observed these two stars.

NGC: None.

There is nothing astronomically significant about M40, except that it is a historical curiosity. In 1660 John Hevelius reported "a nebula above the back" of Ursa Major. This nebula is now widely accepted to be nothing more than a double star that must have appeared nebulous in Hevelius's "old, imperfect instrument," as Flammarion pointed out. Messier, too, reported trying to find a nebula at the coordinates given by Hevelius, but identified only two stars. Although he was not fooled by them, for some reason he still decided to include the pair in his catalogue. The nebulous appearance of a close double should not be surprising to anyone who has swept the Milky Way with binoculars or a rich-field telescope. Close pairings of stars often look like comets, especially at low magnifications.

Ironically, Flammarion based his conclusion on observations of a different double than the one Messier encountered in his search for the Hevelius nebula. And Smyth did the same! Confusing the matter further is that, when precessed, the position of Hevelius's nebula coincides with the position of the star 74 Ursa Majoris, which is not even double!

Regardless, Mallas verified that the object Messier observed and catalogued is Winnecke 4, a double star rediscovered in 1863 by A. Winnecke at Pulkova Observatory in Russia. And because that double is the one recorded in the Messier catalogue, it is the one we will locate.

Winnecke 4 is easily found about ½° northeast of 70 Ursa Majoris, near Delta (δ) in the bowl of the Big Dipper. Its stars have magnitudes of 9.0 and 9.6, and are separated by 49″. At 23×, Delta, 70, 74, and 75 Ursae Majoris, and M40 all fit in the same low-power field. The barred-spiral galaxy NGC 4290 lies nearby and is easily visible in the 4-inch as an irregu-

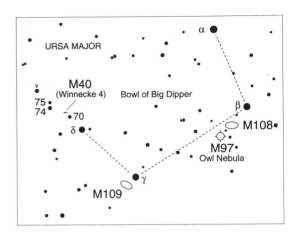

lar splotch. *The Deep-Sky Field Guide to Uranometria 2000.0* lists its magnitude at 11.8, though I estimate it to be 11.5. When I push my vision to the limit, I can see a bar, which looks like a detached segment, almost making the galaxy appear like a double nebula. More challenging and at the limit of visibility is yet another galaxy, NGC 4284, immediately west of NGC 4290. Can you see it? Several times I suspected it. The galaxy's magnitude is listed at 13.5, but it may be brighter.

M41

Little Beehive
NGC 2287
Type: Open Cluster
Con: Canis Major
RA: 6ʰ 46ᵐ.0
Dec: –20° 45′.3
Mag: 4.5
Dia: 40′
Dist: 2,100 l.y.
Disc: John Flamsteed, 1702; Giovanni Batista Hodierna recorded it before 1654, and Aristotle noted it about 325 B.C.

As Canis Major, the Great Dog of Orion, rises above cool winter landscapes, open cluster M41 hangs below its collar like an ice-covered tag reflecting moonlight. One of the faintest objects recorded in classical history, M41 was seen by Aristotle as a star with a tail. With the naked eye, I find it surprisingly similar to the Beehive Cluster (M44), only smaller. Virtually sitting on 6th-magnitude 12 Canis Majoris, about 4° south of Sirius (the brightest star in the night sky at magnitude −1.5), M41 shines at magnitude 4.5 and fills an area of sky 30 percent larger than the full moon. From nineteenth-century England, Webb hailed it as a "superb group visible to the naked eye."

Unfortunately for many of today's observers, the naked-eye view is not as superb, or even possible, thanks to light pollution. But if you do enjoy dark skies, M41 looks to the naked eye like a ghost image of Sirius, or a falling clump of snow illuminated by Sirius's fire-fused crystal blaze. In 7×35 binoculars, both Sirius and M41 fit in the field of view, and binoculars resolve the cluster quite well. But even with the unaided eye I get the distinct impression of resolution – a provocative sight, because of the 80 or so stars in this 100-million-year-old cluster, about 50 shine between 7th and 13th magnitude; at least three of them are 7th magnitude, and seven others are 8th magnitude. I think that after several observations over time, an astute observer could resolve some individual stars with the unaided eye. The stars in M41 are spread across about 25 light years, and the cluster's central density is a loose 6 stars per cubic parsec.

The first thing I notice in the telescope is a pair of equally bright red stars in the center of the cluster. But this observation is rather enigmatic, considering that most accounts refer to a *single* red star there. Webb saw "larger stars in a curve with ruddy star near centre." Jones described a "central, slightly orange star," and Mallas referred to it as the "famous red star at the center." Burnham notes that the "central reddish star" is a Type-*K* giant, and that several other Type-*K* giants are known in the cluster. The central star may also be variable. I compared my drawing of M41 (with two central red stars) with the drawing by Jones (with one central red star) in *Messier's Nebulae and Star Clusters*. Interestingly, I can identify my second red star due east of Jones's single red star. But he drew it markedly fainter!

Curiously, Smyth writes of M41: "A double star in a scattered cluster.... *A*9, lucid white, *B*10, pale white." He could have been referring to a double star on the western edge of the cluster, as Jones believed. But if so, then the prosaic Smyth, who often refers to spectacularly colored stars in clusters, apparently did not notice any predominantly red stars in M41.

Using my imagination, I can see the pattern of bright stars in M41 outlining a fruit bat, or flying fox. The bat's wings are slightly opened in a curve. The bat is hanging by a branch (which flying foxes do), and 12 Canis Majoris makes a tempting piece of fruit for the bat to eye. This pattern can also be made out in 7×35 binoculars. At 72× the entire bat asterism fits perfectly in the field of view. But because of its larger apparent size, the

MESSIER: [Observed 16 January 1765] Cluster of stars below Sirius, close to ρ Canis Majoris. This cluster appears nebulous in a simple one-foot refractor. It is no more than a cluster of faint stars.

NGC: Cluster, very large, bright, little compressed, stars of 8th magnitude and fainter. (*Note:* The *New General Catalogue* incorrectly recorded this cluster as M14.)

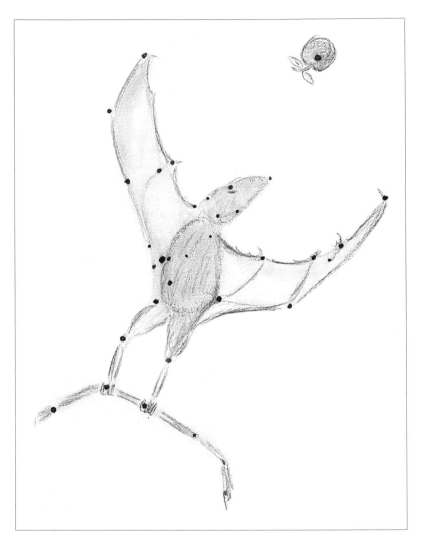

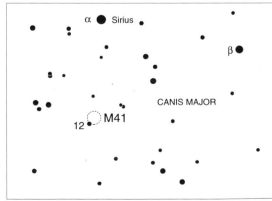

Deep-Sky Companions: The Messier Objects

asterism loses some of its impact. The stars do stand out more boldly though, and the dark rifts between the stars seem to slither across like blacksnakes. The cluster suddenly appears loosely gathered into groups of stars that form triangles, arcs, lines, and other geometrical patterns, as if in a cataclysmic explosion, with celestial debris flying off in various directions.

M42

The Great Orion Nebula
NGC 1976
Type: Emission Nebula and Cluster
Con: Orion
RA: 5^h 35^m.3 (Trapezium)
Dec: –5° 23′ (Trapezium)
Mag: 3.7
Dim: 1°.5 × 1.0°
Dist: 1,500 l.y.
Disc: Nicholas Peiresc, 1611

MESSIER: [Observed 4 March 1769] The position of the beautiful nebula in Orion's sword, around the star θ, which lies within it together with three other, fainter stars, which can be seen only with good instruments. M. Messier went into great detail about this great nebula; he gives a drawing, made with the greatest care, which may be found in the *Mémoires de l'Académie 1771*, plate VIII. Huygens discovered it in 1656, and many astronomers have observed it subsequently. Plotted in the English *Atlas Céleste*.

NGC: Magnificent Theta Orionis and the Great Nebula.

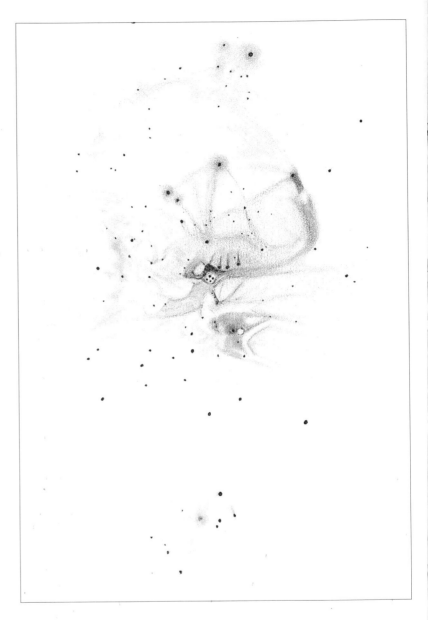

One of the greatest paradoxes in visual astronomy is that Galileo (who paid great attention to Orion, according to Webb) apparently never noticed the Great Nebula. Furthermore, there appears to be no mention of it in medieval records. Yet here is one of the grandest naked-eye nebulae in the heavens, dangling from one of the best-known asterisms (Orion's Belt) in one of the most famous and brilliant constellations. That

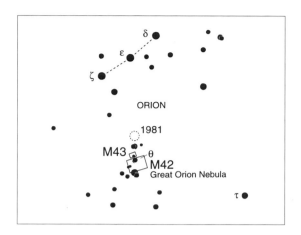

Galileo would have missed seeing it is especially puzzling given that Al-Sufi and others had noticed the fainter "nebula" in Andromeda (M31) as early as A.D. 905! Messier also miscredits Huygens with the discovery of the Orion Nebula, apparently unaware of the earlier observations made by Peiresc and others. But even Huygens's words should alarm modern-day observers: "There is one phenomenon among the fixed stars worthy of mention, which as far as I know, has hitherto been unnoticed by no one and indeed, *cannot be well observed except with large telescopes.*" (Emphasis is mine.)

Yet, in how many New England winters have I looked across snow-laden fields to see this nebula, in the middle of Orion's sword, looking like angel's breath against a frosted sky. Even today, M42 is visible with the naked-eye from the heart of downtown Boston, Massachusetts!

The Orion Nebula is an enormous cloud of fluorescent gas, predominantly hydrogen – with traces of helium, carbon, nitrogen, and oxygen – 40 light years in diameter. It glows by the ultraviolet radiation streaming from Theta (θ) Orionis, a bright grouping of four massive stars, whose light energy is burning a hole through the mackerel clouds of the nebula's sharply defined interior. Commonly called the Trapezium, these four jewels shine between 5th and 8th magnitude, and it is the brightest member that produces 99 percent of the energy that illuminates the clouds of gas we see through our telescopes. All the hot, young Trapezium stars probably began shining a mere 1 million years ago (our sun is 4.5 billion years old). And there are about a thousand unseen members (300,000 to 1 million years old) hiding in the dense cloudscapes. Recent Hubble Space Telescope images show the expansive nebula in unprecedented detail: tumultuous swirls of colorful churning gas; 40-billion-mile-long "comets" of dust and gas, whose comas enshroud newborn stars; dark protoplanetary disks silhouetted against the nebula's spectral

M43

NGC 1982
Type: Emission Nebula
Con: Orion
RA: 5^h 35^m.6
Dec: –5° 16'
Mag: 6.8
Dim: 20' × 15'
Dist: 1,500 l.y.
Disc: Jean-Jacques Dortous de Mairan, before 1750

MESSIER: [Observed 4 March 1769] Position of the faint star that is surrounded by nebulosity and that lies below the nebula [viewed with south up] in the sword of Orion. M. Messier depicted it on the drawing of the large nebula [M42].

NGC: Remarkable, very bright and large, round with a tail, much brighter in the middle, contains a star of magnitude 8 or 9.

glow. The HST has confirmed what astronomers have long suspected: M42 has all the ingredients for solar and planetary creation.

At 23×, I have no trouble seeing the overall shape of the nebula as a blazing comet, the kind that invoked fear and superstition in the impressionable population of days gone by. Leading to this cometary image is the broad sweeping gaseous mass to the southwest and the curved jets streaming from the fractured nucleus. (Remember Comet West and how its nucleus split into four fragments when it rounded the sun in 1976?) Low power is also required to follow the large bubble of vapors rising above that turbulent central region; the thick northern arms of the bubble are the bat wings E. E. Barnard saw beating against the darkness to the south. The overwhelming structural detail of this nebula makes it appear larger than it is. M42 is 1°.5 at its widest, which is three times smaller than M31, the Andromeda Galaxy, and roughly the same size as open cluster M44. The entire Orion Nebula, however, fits nicely in most low-power eyepieces.

How can anyone draw the Orion Nebula and do it justice? The inner square of nebulosity surrounding the Trapezium is a chaotic witch's brew. This stellar grouping, under very steady conditions, looks like an emerging embryo cradled in a soft womb of nebulosity. The Trapezium actually contains several visible fainter stars in and around it. Two of them are commonly seen by amateurs with moderate-size instruments. Use the accompanying chart to search for these more difficult stars. The easiest is the 11th-magnitude star "E," about 4" due north of "A." Also shining at 11th magnitude is star "F," 4" southeast of "C," but I find this a more difficult object and believe it to be fainter than "E." Burnham writes that both "E" and "F" have been seen with apertures smaller than 3 inches! The other two members shine at 16th magnitude and are challenges for the largest amateur instruments. Texas deep-sky wizard Barbara Wilson and I glimpsed "G" through a 36-inch Dobsonian at the 1994 Winter Star Party, though "H" remained elusive. The "X" on the Trapezium chart marks the location of a star that E. E. Barnard saw at the limit of vision through the 36-inch refractor; he saw it on two nights in 1888 and 1889, but to my knowledge it has not been recorded visually by anyone since.

Look carefully at the inner square at high magnification and try to locate the four sharp flares shooting off to the south of it and the large, hedgerow "prominence" to the southwest. Now study the dark bay northeast of the Trapezium (Smyth called this the "fish's mouth"); use averted vision and you should see several herringbone bridges crossing it. If you concentrate solely on the dark bay and forget the nebulosity, a powerfully dark image appears. It is a mighty silhouette and you can follow its course throughout the entire region. The faint nebulosity forming the bay's

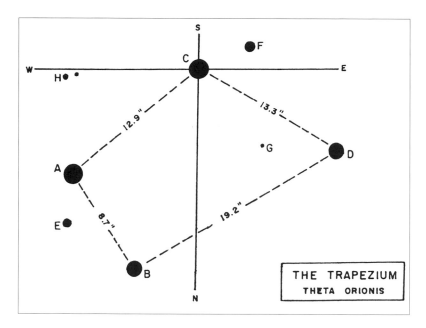

THE TRAPEZIUM
THETA ORIONIS

northern bank appears to have turbulent edges near the Trapezium. And its pale gray color and soft texture are so unlike the stronger green wreath-like nebulosity in the main body of M42. The Orion Nebula is one of the few gaseous clouds that shows color through amateur telescopes. Photographers have no problem recording vivid reds and greens in it with proper exposures. Of course, Hubble images depict a full spectrum of vibrant colors throughout the nebula. Our eyes, on the other hand, are largely insensitive to color at night. So visual observers have to settle for subtle shades of green or red. Interestingly, when we do see these colors, pale as they might be, our minds tend to exaggerate their intensity, and we say the color is "strong," as I did above.

M43 is a wedge of nebulosity surrounding a 7th-magnitude star (called Bond's star) 10′ northeast of the Trapezium. Commonly overlooked because of its proximity and association with the Orion Nebula, M43 harbors a wealth of detail that is well worth becoming acquainted with, so plan an evening with it. You will have to block out the commanding presence of its larger neighbor, but one way to seclude M43 is to use high magnification. See if you can resolve the tiny dark pool due east of Bond's star. One night a short but strong earthquake struck while I was examining M43. The telescope began to shake and, when it did, my eye caught sight of some extremely faint extensions looping to the west and a long wedge of material flowing to the east. Instead of waiting for an earthquake, try tapping your telescope tube and see if you can make out these subtle features. Also, be sure to move north of M43 to NGC 1977, more nebulosity

M44

whose wavy texture is delightful at low powers. Then move farther north to the loose open cluster NGC 1981, to complete your tour of Orion's sword.

By the way, returning to M42 and the subject of challenges, I have also spent considerable time peering at the region just inside the Great Nebula's arcing western rim and searching for "GOD" – a series of dark nebulous swirls that spell out the word GOD in capital letters. Although I have seen portions of these letters, I have yet to piece together the details in a single strong view. This challenge might be a bit too difficult for the 4-inch, but I do urge those with larger telescopes to look for GOD in the Orion Nebula.

M44

Praesepe or *Beehive Cluster*
NGC 2632
Type: Open Cluster
Con: Cancer
RA: 8^h 40^m.4
Dec: + 19° 40′
Mag: 3.1
Dia: 1°.2
Dist: 515 l.y.
Disc: Known since antiquity

MESSIER: [Observed 4 March 1769] Cluster of stars known as the nebula in Cancer [Praesepe]. The position given is that of star C.

NGC: None.

Cancer is the only constellation whose brightest stars are fainter than a Messier object within its boundaries. In fact, if it weren't for the mystifying cloudy appearance of open cluster M44, which draws your gaze to the surrounding 4th-magnitude stars, it is conceivable that dim Cancer might have either gone unnoticed by ancient stargazers or have been envisioned differently. To the naked eye, the 3rd-magnitude glow of M44 looks like the bearded head of a tail-less comet passing between the 4th-magnitude stars Gamma (γ) and Delta (δ) Cancri.

Popularly known as the Beehive Cluster or Praesepe (The Manger), the misty glow of M44 had a more macabre significance in ancient China, where it was seen as Tseih She Ke, "exhalation of piled-up corpses." M44 did not go unnoticed by Galileo, who described its telescopic appearance

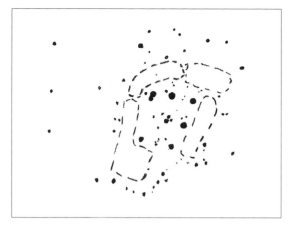

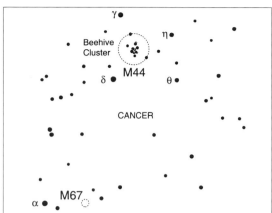

as "not one only but a mass of more than 40 small stars." In modern tele-scopes, about 200 cluster members ranging from magnitude 6 to 14 are visible, 80 of which are brighter than magnitude 10. The age of the cluster is estimated to be 400 million years, and its core measures 11 light years in diameter.

At first glance, M44 appears round to the naked eye, but if you con-centrate, it should look elongated roughly north–south. You might also begin to suspect arms extending in various directions and stars shining within the central haze. Is M44 resolvable to the unaided eye? Yes, and it's not difficult to do! In fact, the cluster contains 15 stars between magni-tudes 6.3 and 7.5. One night, while observing at 14,000 feet, I gazed at the Beehive without optical aid while occasionally breathing oxygen through an oxygen mask. To my amazement, the haze of unresolved stars vanished and the individual cluster members stood out boldly as rock-steady pearls, each tiny, bright disk surrounded by an intense black rim. I did not count the stars that night because they seemed so numerous. I used the oxygen in brief spurts to help keep me alert, because, at 14,000 feet, the atmosphere is 40 percent thinner than at sea level. Lack of oxygen can also decrease the sensitivity of the eyes. For example, without supplemental oxygen at 14,000 feet, I had to use averted vision to see the Andromeda Galaxy well; 10,000 feet lower, however, the galaxy was an incredible naked-eye sight with direct vision. There is no reason for sea-level observers to use additional oxygen, because you're working in an oxygen-rich environment (hyperventilating, however, can increase the immedi-ate supply); too much oxygen will make you sick. I did count the stars in M44 from 9,000 feet and 4,000 feet without oxygen. With a long, dedicated effort, I recorded a dozen cluster members without optical aid – that's twice the number of stars most observers *casually* see in the Pleiades (M45)!

Through the 4-inch at 23×, the entire 1°.2-diameter cluster is visible, and the stars seem to swarm with sparkling madness – truly a beehive of nervous starlight. At least one-fifth of all the stars are doubles! The main body of stars is sharply outlined by several strong dark lanes (reminiscent of the naked-eye view from 14,000 feet) with a tiny coalsack to the north-west. High-power or large telescopes will bring out the faint stars within the dark lanes, but why destroy such a visually stunning illusion of emptiness?

I would be curious to know how the celestial backdrop affects brightness estimates of open clusters in the Milky Way band. For example, one reason M44 appears so visually distinct is that its brilliant orb of starlight is seen against the star-poor region of Cancer, which is off the main stream of the Milky Way. It is seen projected against a darker background than, say, either M6 or M7, which are nestled in the most brilliant region of the Milky Way. I can only imagine how stunningly bright M7 would appear if it were seen against the stars of Cancer and how dim M44 would appear if seen against the heart of the Milky Way!

M45

Pleiades or *Seven Sisters*
NGC (not listed)
Type: Open Cluster
Con: Taurus
RA: 3^h 47^m.5 (Alcyone)
Dec: + 24° 06′.3 (Alcyone)
Mag: 1.5
Dia: 2°
Dist: 407 l.y.
Disc: Known since antiquity

MESSIER: [Observed 4 March 1769] Star cluster known as the Pleiades. The position given is that of the star Alcyone.

NGC: None.

On crisp winter evenings, this brilliant 1st-magnitude open cluster rides high on the shoulder of Taurus the Bull, whose V-shaped face itself (minus Aldebaran) is another, brighter open cluster – the Hyades. Together these clusters are among the most alluring sights in the heavens. But, the larger, more open Hyades pales in comparison to magnificent M45. With the naked eye, the Pleiades looks like a tiny dipper, a forest of starlight bathed in moonlit mist, or a distant gathering of veiled brides. Indeed, photographs show the entire cluster swaddled in ice-blue nebulosity, which reflects the light of these young, hot stars. The stars of the Pleiades emerged from their dusty cocoon some 20 million years ago. And some astronomers have conjectured that rather than the surrounding nebula belonging to them, the stars may just be passing through a nebulous region in Taurus.

Just how many stars (Pleiads) in the Pleiades are truly visible with the naked eye is the subject of some debate. Traditionally, the number has been seven. But that count stems from ancient Greek mythology and refers to the seven doves that carried ambrosia to the infant Zeus, or to the seven sisters who were placed in the heavens so that they might forget their grief over the fate of their father, Atlas, condemned to support the sky on his shoulders. Although largely symbolic, the age-old association of the Pleiades with the number seven remains fixed to this day – to the point that some observers swear they cannot see more than seven members, even though the Pleiades contains *10 stars brighter than 6th magnitude*. Some observers question how it is possible to see 10 Pleiads in the Seven Sisters (a demonstration of the power of words or, as Clerke inferred, the power of the Pleiades over human affairs.) The fact is that almost *three times* that magic number of stars can be seen without magnification by an astute observer under dark skies. Clerke notes that Kepler's tutor, Maestlin, saw 14, and he mapped 11 before the invention of the telescope. Archinal reports routinely seeing 12 Pleiads and sometimes 14 under good skies. Houston counted 18, and about 20 years ago I logged 17 from Cambridge, Massachusetts! The trick is to spend a *lot* of time looking and plotting.

Another lingering debate concerns whether the nebulosity associated with the cluster can be seen with the naked eye. Some argue that the haze is just an illusion, a vision created by the tight gathering of stars whose unresolved light simply *appears* fuzzy – just as close doubles do at low power (like M40). Others, however, claim that the cluster appears just as Tennyson described it in *Locksley Hall:* "Many a night I saw the Pleiades, / rising thro' the mellow shade, / glitter like a swarm of fireflies / tangled in a silver braid." Houston once scoffed, "Why does everyone fuss over the Pleiades nebulosity? Of course you can see it with the naked eye!" He went on to say something colorful about the naysayers. I did not reply, and he didn't expect a response; he knew he was preaching to the choir. However, Skiff points out that some of the perceived nebulosity may be scattered starlight. He suggests testing this hypothesis by screening Alcyone, the brightest Pleiad, behind the edge of a building. Try it.

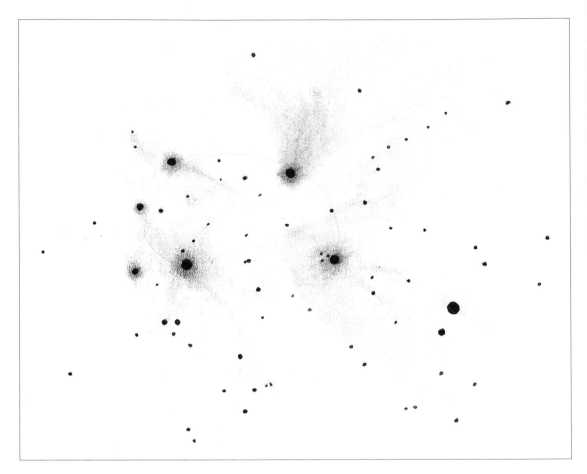

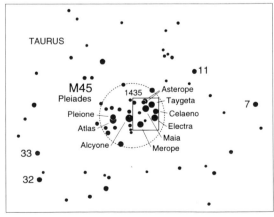

The 70-million-year-old cluster contains about 100 stars in a sphere 14 light years in diameter. It is sailing through space at about 25 miles per second, and it will take them about 30,000 years to move an apparent distance of one moon diameter. The brightest Pleiads are all rapidly rotating, with Pleione spinning about 100 times faster than our sun. And Alcyone, the centralmost and brightest Pleiad, is a thousand times more luminous that the sun.

Through a telescope the Pleiades and its attendant nebulosity fill a low-power field, where they are best seen. The nine brightest members fit well in a 1° field (a true diameter of 7 light years). From dark skies, the cluster looks like a cobwebbed coffin filled with glistening jewels. The individual Pleiads are interconnected by gauzelike veils of nebulosity, the brightest of which surrounds the star Merope (23 Tauri), and is called the Merope Nebula (IC 349).

Just south of Merope is a condensation of nebulosity, NGC 1435, also known as "Tempel's Nebula," which was first noticed in 1859 by W. Tempel with a 4-inch refractor. He described it as a faint stain of fog, like a "breath on a mirror." In 1874 Lewis Swift saw it in a 2-inch refractor at 25×. One century later Houston used an 8-inch reflector from southern Arizona to see the field laced from edge to edge with bright wreaths of delicately structured nebulosity. Interestingly, the great double-star observer S. W. Burnham could not see this nebulosity with the great 18-inch refractor at Dearborn Observatory in Illinois. But faint, diffuse glows can evade large-aperture telescopes.

At first glance the Merope Nebula network looks like a tapered comet tail. But if you take the time to sweep the telescope back and forth over its surroundings, you might notice that this patch is but part of a larger fan of material that sweeps westward. Using low magnification, just stare at the cluster for a while and occasionally tap the telescope tube. You should start to see wide dark lanes running amid the nebulosity, especially at the cluster's center, where it drapes around a bridal veil of nebulosity adorning Alcyone. Study the drawing and photograph and look for prominent streaks of nebulosity, especially around Maia and Electra.

M46

NGC 2437
Type: Open Cluster
Con: Puppis
RA: 7^h 41^m.8
Dec: −14° 49′
Mag: 6.1
Dia: 20′
Dist: 5,300 l.y.
Disc: Messier, 1771

MESSIER: [Observed 19 February 1771] Cluster of very faint stars between the head of Canis Major and the two rear hoofs of Monoceros, determined by comparing the cluster with the sixth-magnitude star Flamsteed 2 in Argo Navis [now 2 Puppis]. These stars can be seen only with a good telescope. The cluster contains some nebulosity.

NGC: Remarkable cluster, very rich, bright, and large, involving a planetary nebula.

Use your binoculars and scan about 15° (a fist- and two finger-widths) east of Sirius in a line due south of Procyon and Alpha (α) Monocerotis. There, hiding among the monotonously faint star fields of the Puppis Milky Way, are the curious Messier open clusters M46 and M47. They lie a mere 1½° (a finger-width) apart but look markedly different. The farther east of the two, M46, appears like a round, uniform 6th-magnitude glow, while gawky M47 is an irregular gathering of reasonably bright but dissimilar stars. And though both clusters have impressive apparent diameters three-quarters the size of the full moon, their wildly different appearances can distract you from that realization. What is immediately obvious, however, even in binoculars, is that whatever M46 lacks in visual grandeur, M47 lacks in visual grace, and vice versa. It's like trying to compare a flower with a rock.

At 23×, M46 is a bright sphere of finely resolved starlight with faint, clockwise spiral structure. If you sweep the telescope to the west, you might notice that M46 appears to be on the eastern end of a wide, V-shaped string of 9th-magnitude stars connecting it to M47. But this is deceptive: M46 is 3,750 light years more distant than M47 and is actually somewhat larger than its closer neighbor! M46 contains 186 stars ranging from magnitude 10 to 13 distributed rather evenly over an area of 20′ with a loose central gathering. But this central gathering is also somewhat illusory, because moderate to high magnification will reveal that M46 has a dark, not a bright, core. With imagination, I see that void as an atoll, with a 2′-diameter black lake rimmed by 10th-magnitude coral specks that break a wave of stars flowing around it from the south.

There is yet another illusion associated with M46. It appears to contain a tiny planetary nebula, NGC 2438, which is distinctly visible in the accompanying photograph. But the cluster and nebula are not physically associated, because the cluster is 5,300 light years distant whereas the nebula is 6,520 light years away. Positioned just a few arc minutes north of the cluster's center, this 11th-magnitude planetary measures only about 1' in diameter. I suspected it at 23×, but 72× shows it clearly as a ghostly mote among the multitude. With averted vision at high power, it looks like the glow from the flame of a distant candle. Try, if you must, for the planetary's central star, but the task might be daunting, given that it shines at about 16th magnitude! Finally, turn your gaze north of the planetary. Do you see the wide canal in the cluster's outer halo of faint stars? It runs southwest to northeast, and its southern bank abuts the planetary.

M47 was once considered a "missing" Messier object. The position that Messier recorded for this fine object was incorrect; no such object exists at his coordinates! Not until 1934 did Oswald Thomas identify M47 as NGC 2422. Twenty-five years later T. F. Morris traced the transcription error that Messier probably made in entering the object's position.

M47 is an elaborate mix of bright and faint stars, which look like they've been tossed together by someone in great haste. The sight makes me feel as if I have opened yet another treasure chest and am looking into a tangle of secret riches. It contains at least 117 members, some as bright as 5th and 6th magnitude, in an area 25' in diameter. Thus, M47 measures a mere 14 light years in diameter, less than half that of M46. I resolved "three" stars in M47 with the unaided eye. I use quotation marks here because the central one is actually the double star Struve 1121, which consists of two opalescent 7.9-magnitude stars separated by 7".4. Three nicely spaced doubles reside in the cluster, as do two 8th-magnitude orange

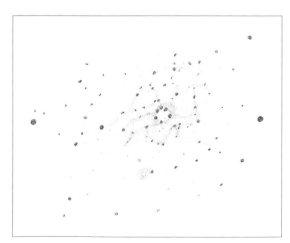

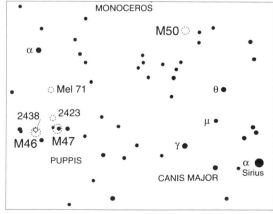

M47

NGC 2422 = NGC 2478
Type: Open Cluster
Con: Puppis
RA: 7^h 36^m.6
Dec: −14° 29′.0
Mag: 4.4; 5.7 (O'Meara)
Dia: 25′
Dist: 1,550 l.y.
Disc: Messier, 1771; recorded
by Giovanni Batista Hodierna
before 1654

MESSIER: [Observed 19
February 1771] Cluster of
stars not far from the previous
one [M46]. The stars are
brighter. The center of the
cluster was determined
relative to the same star,
Flamsteed 2 Argo Navis [now
2 Puppis]. The cluster
contains no nebulosity.

NGC: Cluster, bright, very
large, pretty rich, with large
and small [bright and faint]
stars.

stars (one north of the brightest star, the other to the south). I am quite surprised at my magnitude estimate of 5.7, which is fainter by more than a magnitude from that in Luginbuhl and Skiff's *Observing Handbook,* and by 0.5 magnitude from the one offered by Jones.

Return to low power and follow a chain of 8th- to 9th-magnitude stars off M47 north to the smaller, 6.7-magnitude NGC 2423, then look to

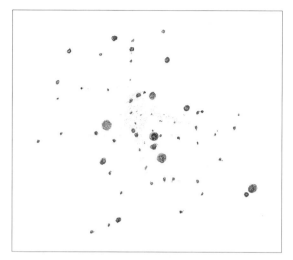

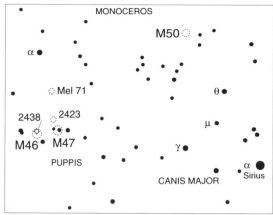

the northwest and you should see another chain of 8th-magnitude and brighter stars curving to yet another, still smaller patch of stars: open cluster Melotte 71. Together all these clusters and strings form a pattern that reminds me of the alien craft depicted in the 1953 film *The War of the Worlds.*

M48

NGC 2548
Type: Open Cluster
Con: Hydra
RA: 8^h 13^m.8
Dec: –5° 45′.0
Mag: 5.8
Dia: 30′
Dist: 2,000 l.y.
Disc: Messier, 1771

MESSIER: [Observed 19 February 1771] Cluster of very faint stars, without nebulosity. This cluster is close to three stars that lie at the root of the tail of Monoceros.

NGC: Cluster, very large, pretty rich, pretty much compressed toward the middle, 9th to 13th magnitude stars.

The most intriguing "missing" Messier object, M48 is now believed to be NGC 2548 – a knitted stellar gathering of about 80 stars between magnitude 8 and 13 located in a rather inconspicuous region of Hydra where it borders Monoceros. M48 marks the southern tip of a nearly equilateral triangle with 3.4-magnitude Epsilon (ε) Hydrae, in the head of the Serpent,

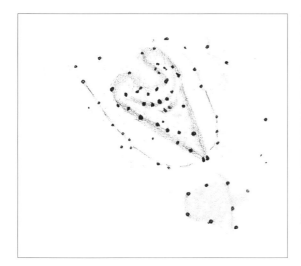

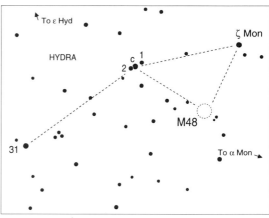

and 0.4-magnitude Alpha (α) Canis Majoris, and lies only 3½° southwest of the obvious naked-eye grouping of 1, 2, and c Hydrae.

Messier's recorded position for M48 is off by about 5° in declination, placing it due north of NGC 2548 (where there are no clusters). Astronomical historian Owen Gingerich, in his chapter in the 1978 classic, *The Messier Album*, explained:

> Messier determined the declination of a nebula or cluster by measuring the difference between the object and a comparison star of known declination. . . . Since no conspicuous star is located 2½° away in declination, we cannot account for this position by another error in sign [as was the case with M47]. . . . Messier did not publish the name of the [comparison] star he used, and his original records are apparently no longer extant. . . . Thus, a careful survey of the region described by Messier leads to the conclusion that NGC 2548 is the cluster that the French observer intended as his 48th object, for lack of any other cluster nearby that fits the description.

Although I agree with Gingerich that NGC 2548 is the most likely candidate for M48, I must also relate the following experience.

When I went out to make my first observation of M48 for this book, I began by looking for the cluster with the unaided eye. I easily located the line of stars 1, 2, and c Hydrae, and then immediately noticed a diffuse patch south of them. I swung the telescope to this region and swept the area at 23×. To my amazement, I could not find the cluster. I looked up again, located the hazy patch, and tried to find the cluster, but without success. Then I realized my mistake, I was looking too far to the southeast. I moved the telescope 5° northwest and – *bang* – there was M48.

I didn't think anything of this experience until I repeated that very same mistake the next time I went out! That's when I noticed *two* naked-eye glows south of 1, 2, and c Hydrae: one is about 3½° southeast of them, the other is about 3½° southwest of them; they are separated by about 5°. Furthermore, M48 is the *fainter* of these two naked-eye patches! There is another geometrical similarity. M48 marks the southern apex of a triangle with c Hydrae and Zeta (ζ) Monocerotis to the northwest, whereas the nameless patch forms the southern apex of a triangle with c Hydrae and 31 Monocerotis to the southeast.

I viewed this mystery cloud through the 4-inch at 23× and found it to be a loose collection of similarly bright stars. When I placed it in my 7× 35 binoculars and slightly defocused the view, I saw a small "fuzzy" object (the unresolved loose gathering of stars in the mystery cloud) surrounded by a brighter circlet of stars (giving the cloud brightness and breadth). Could this curious object have been Messier's real 48th entry?

Through the 4-inch, M48 is a tremendously pleasing cluster, a perfect arrowhead of bright stars with a tight, elliptical, off-axis core. Like Rosse, I find the cluster riddled with dark lanes and openings. One dark feature I call the "keyhole" – it's a notch in the base of the arrow to the southwest. Actually, I like to call M48 the "alligator," because, it reminds me of an aerial view of a stalking alligator (when all one sees is its triangular head and snout). With this image in mind, I see a wake of stars flowing around the snout, and, to the northeast, a circle of stars representing a turtle – an alligator's favorite meal.

M49

NGC 4472
Type: Elliptical Galaxy
Con: Virgo
RA: 12^h 29^m.8
Dec: +8° 00'
Mag: 8.4
Dim: 10'.2 × 8'.3
Dist: 55 million l.y.
Disc: Messier, 1771

MESSIER: [Observed 19 February 1771] Nebula discovered close to the star ρ Virginis. This is rather difficult to see with a simple three-and-a-half-foot refractor. M. Messier compared the comet of 1779 with this cluster on 22 and 23 April: the comet and the nebula had the same luminosity. M. Messier plotted this nebula on the chart showing the path of this comet, which will appear in the Academy volume for the same year, 1779. Observed again 10 April 1781.

NGC: Very bright, large, round, much brighter toward the middle, mottled.

M49 is an enormous elliptical system that lurks at the center of a subcluster of galaxies, called the Virgo Cloud, in the heart of the vast Virgo Cluster – a rich concentration of galaxies spanning some 20 million light years of space. In the early 1990s, ROSAT (an earth-orbiting x-ray observatory) imaged M49 and revealed it to possess a halo of hot (10,000,000 K) gas. Observations with previous x-ray telescopes have shown that the hot gas in galaxy clusters cannot be confined by the combined gravity of the gas and the galaxies alone. Thus, the ROSAT data imply that perhaps 75 to 95 percent of the mass of small galaxy clusters, such as the M49 subcluster, could be in the form of dark matter. It has also been discovered that elliptical galaxies that lie at the center of galaxy clusters lack the raw materials from which new stars form, so the ellipticals are excessively red.

To find M49 among the multitude of galaxies in the Virgo Cloud, look 2½° southwest of 6th-magnitude 20 Virginis. There you'll spy its very compact, 8th-magnitude glow nestled between two 6th-magnitude stars. The galaxy shows up well in binoculars from a dark-sky site.

At moderate power, M49 reveals a bright nucleus surrounded by a tight inner core and a diffuse halo. A 12.5-magnitude star punctuates the

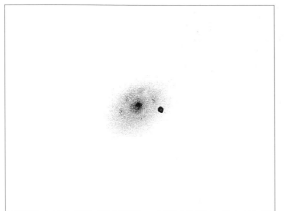

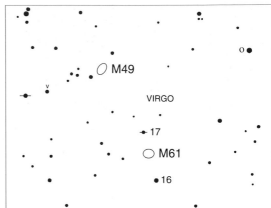

galaxy's bright eastern fringe like a splendid supernova. The overall sight reminds me of an unresolved globular cluster or the head of a comet. Barnabus Oriani, an Italian observer of the eighteenth century, thought the haze looked exactly like the comet of 1779 (apparently, so did Messier).

But look carefully at the nucleus with high power. I get hints of inner structure, namely, a mottled nuclear region and two arcs of light at the outer edge of the inner halo, one to the northeast, the other to the south-west; they also appear mottled and cause me to suspect spiral structure, though this may be an illusion. D'Arrest, however, believed he resolved the misty glow of M49 into stars of 13th to 14th magnitude near the edges. Speaking of illusions, do you see a cross of faint knots or stars bordering the bright inner region? It resembles the gravitational-lensing effect known as Einstein's Cross. The visible outer envelope itself is very deceiving: at high power, it makes the galaxy appear round, but moderate to low magnifications reveal a broad shield of diffuseness that transforms this sphere into an elliptical glow oriented northeast to southwest.

When using the telescope, refer to the detailed finder chart for M49 to scan the area, because, as I quickly discovered, the low-power field is chock full of galaxies, and they're all pleasantly distracting. The most prominent of them is NGC 4526, another elliptical about 1° southeast of M49, between two 7th-magnitude stars.

M50

NGC 2323
Type: Open Cluster
Con: Monoceros
RA: 7^h 02^m.7
Dec: − 8° 23′
Mag: 5.9
Dia: 15′
Dist: 3,300 l.y.
Disc: Probably Giovanni
Domenico Cassini, before
     1711; Messier
rediscovered it in 1772

MESSIER: [Observed 5 April 1772] Cluster of faint stars of differing brightness, below the right thigh of Monoceros, above the star (in the ear of Canis Major, and close to a seventh-magnitude star. It was while observing the comet of 1772 that M. Messier observed this cluster. It has been plotted on the chart for that comet that he prepared, *Mémoires de l'Académie 1772*.

NGC: Remarkable cluster, very large, rich, pretty compressed, elongated. The stars range from 12th to 16th magnitude.

M50, an obscure open cluster in an equally obscure constellation (Monoceros), has an aura of enigma about it. Giovanni Domenico Cassini, father of Jean Dominique Cassini (who discovered the first gap in Saturn's rings in 1675), appears to have chanced upon this object before 1711. But this fact was only related by his son in his *Elements of Astronomy*, where he mentions his father sighting a nebula in the area between Canis Major and Canis Minor. Messier looked for the object in 1771 but did not find it, concluding that the elder Cassini had probably witnessed a passing comet. One year later, however, Messier did sweep up an open cluster in Monoceros that was to become the 50th entry in his catalogue.

You can use binoculars to locate M50. It will appear as a bright 6th-magnitude glow bathed in a rich Milky Way field just 7° north of Gamma (γ) Canis Majoris – a 4th-magnitude star about 5° (two finger-widths) east of Sirius. Once you've found this snowy blur, lower the binoculars and try to see it with the naked eye. This might be a challenge if your skies are at all light polluted.

At low power the cluster's true extent is rather hard to make out,

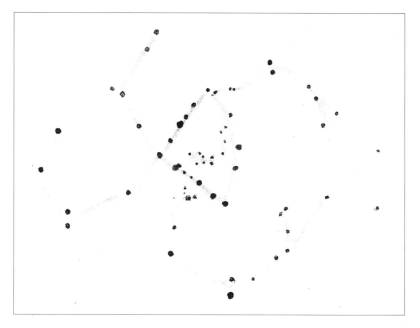

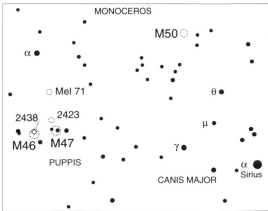

because the Milky Way surrounding M50 is bursting with stellar clumps and asterisms. Indeed, Skiff says it is hard even to determine M50's declination because of irregular concentrations of stars in the immediate vicinity. Wallenquist found at least 50 members in the central 14′ of the cluster, and Skiff counts 150 members across 25′ in a 10-inch telescope. Archinal tallies only 80 cluster members in a 14′ diameter, though. The question is, Where does the cluster really end?

With moderate magnification, the main pattern of stars forms a crude cross whose broken arms extend nearly 1° on a side. These prominent but star-poor arms radiate from the cluster's bleak center; they would make a striking spiral pattern, except that one of the arms is "turning" the wrong way. But, again, are these long stellar extensions part of the true cluster? If you cut back to low power and look at the cluster's overall shape, you will see fainter stars filling in the gaps between the arms, giving the cluster a distinctly oval appearance, though it is just slightly out of round. M50's shape reminds me of Omega Centauri's.

Now concentrate on the heart of the cluster. Do you see a distinct right-angle pattern of stars with a curved hypotenuse to the east? A smaller cluster of stars resides in the northern section of that warped right triangle. This minicluster contains about a dozen members that form a scooplike asterism. Apparently, Luginbuhl and Skiff noticed this grouping too, observing that "the middle of the cluster is relatively empty except for one small group." They also recorded a dark patch near the cluster's center, which is probably why M50's heart looks so empty. The stars of M50 are slightly reddened, implying some obscuration by dust along our line of sight.

Many observers over the past century and earlier have commented on a red star in the southern portion of the cluster; Smyth, d'Arrest, and Webb all noted it. But I did not see it, nor did Mallas, who wrote, "A red star mentioned by Admiral Smyth and T. W. Webb is presumably the 8th-magnitude object about 7′ south of the cluster's center, but the color was not conspicuous to [me]." Indeed, Skiff notes that photometry of this magnitude-7.8 star on the southern edge shows it to be definitely red. Could the discrepancy be because this star is a variable? Or, does it just reflect individual differences in color perception of faint objects? Archinal viewed the star with a 20-inch f/7 reflector and described it as "particularly bright orange." He suspects the reported variations in the star's color are due to variations in observers' eyes, particularly at low light levels.

Use low power to follow a meandering river of 9th- to 10th-magnitude stars flowing south of M50. It is even more striking in binoculars, looking like a fuzzy strand of DNA. It winds its way toward another weak and loose open cluster, NGC 2335, 2° to the southwest.

M51

Whirlpool Galaxy
NGC 5194
Type: Spiral Galaxy
Con: Canes Venatici
RA: 13ʰ 29ᵐ.9
Dec: +47° 12'
Mag: 8.4
SB: 13.1
Dim: 11'.2 × 6'.9
Dist: 15 million l.y.
Disc: Messier, 1773

MESSIER: [Observed 11 January 1774] Very faint nebula without stars, near the more northerly ear of Canes Venatici, below the second-magnitude star (in the tail of Ursa Major. M. Messier discovered this nebula on 13 October 1773, when observing the comet that appeared in that year. It may be seen only with difficulty with a simple three-and-a-half-foot refractor. Nearby there is an eighth-magnitude star. M. Messier plotted its position on the chart of the comet observed in 1773 and 1774, *Mémoires de l'Académie 1774*, plate III. It is double: both parts have bright centers, and they are 4'35" apart. The two atmospheres are in contact; one is much fainter than the other. Observed several times.

NGC: A magnificent object, great spiral nebula.

Just as the Crab Nebula, M1, caused Messier to undertake his catalogue, the Whirlpool Galaxy, M51, caused me to undertake my own version of his catalogue. My reason for starting, though, was different from Messier's. He would have preferred that you avoid the objects in his list, whereas I'm encouraging you to seek them out!

Once believed to be a great swirling nebula, M51 is now known to be the finest example of a face-on spiral galaxy. A near neighbor of our own galaxy, just 15 million light years away, this graceful pinwheel of stars, dust, and gas measures about 50,000 light years across and shines with the luminosity of about 10 billion suns. Rosse detected the nebula's "spiral convulsions" in 1845 with his 72-inch speculum mirror reflector, making M51 the first galaxy shown to have spiral structure. Some eighteenth- and nineteenth-century observers, such as John Herschel, were nearly clairvoyant about the object's nature. Herschel's drawing of M51 shows a split ring surrounding a central condensation, a view that closely resembled his concept of the Milky Way's structure. Smyth went so far as to describe the nebula as a "stellar universe, similar to that to which we belong, whose vast amplitudes are in no doubt peopled with countless numbers of percipient beings." But these statements were based on the knowledge of the day; the astronomers did not envision a galaxy of countless suns but rather a hazy vortex that verified the cosmology established by the French mathematician Pierre Simon de Laplace, who postulated that our solar system condensed out of a rotating gaseous nebula. This notion of M51 being a solar system in formation was not shattered until 1923, when astronomers discovered the true nature of the mysterious spiral nebulae.

M51 is easily spotted in binoculars, glowing at 8th magnitude 3½° southwest of Eta (η) Ursa Majoris, the end of the Big Dipper's handle. As my drawing indicates, M51 does not reveal much detail in small telescopes. But the features I did record are impressive for a 4-inch telescope. For example, the galaxy's main spiral arms are often stated as being difficult for telescopes smaller than 10 inches. Much depends on a practiced eye and very favorable viewing conditions. Still, had I been an observer in the nineteenth century, armed with the telescope I own today, I would not have shied away from at least suggesting the possibility that M51 had spiral structure. The brightest portions of the arms do appear mottled and are clearly separated from the stellar nucleus and its tightly wound core by dusky patches (this is the "ring" Herschel and others saw). I would have agreed with Smyth, who saw M51 as "resembling the ghost of Saturn with its rings in a vertical position."

Messier saw M51 as a double nebula whose "atmospheres" were touching one another. Why, then, didn't he catalogue each bright component separately, as he did for M42 and M43, and for M31 and M32? Perhaps because of M51's small apparent size compared to the mighty Orion Nebula or Andromeda Galaxy. Or was it because the Whirlpool and what is now known to be its companion galaxy, 9.5-magnitude NGC 5195, 4′ to the north, appeared more like a single entity than two distinct ones, perhaps because of an unsteady atmosphere.

Admittedly, the most surprising detail in my drawing is the dim bridge of gas connecting M51 to NGC 5195. The visibility of this bridge in small apertures has long been debated. Some argue it cannot be seen except from under a dark sky and with at least a 12-inch telescope. Others say it can be seen in a 4-inch. Could the view of the bridge in small apertures be illusory – a manifestation created by the proximity of the two galaxies coupled with the knowledge that the bridge exists? I did an experiment to find out.

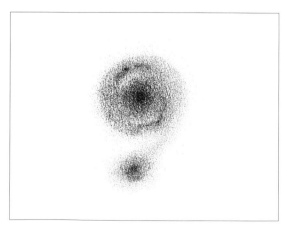

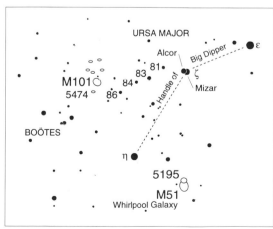

As I mentioned in chapter 2, the nineteenth-century astronomer George Bond recorded the outer extensions of diffuse objects by placing the object outside the field of view and then letting it drift back in until he detected a change in the brightness of the sky background. I tried a modified version of this technique with M51. Using 130×, I moved the telescope toward M51 from the west and found a clear gap between M51 and its companion, until I reached the eastern boundaries of the two, where I encountered a slight increase in brightness that ended the canal. I repeated this sweep several times. Each time, soon after my mental canoe went sailing down that dark canal it collided with this phantom bridge. Then, I reversed the sweep, moving from east to west toward the two galaxies. There was no "entrance" to the canal; I was stopped immediately by the bridge. Try this sweeping technique yourself to discover your own truth about the visibility of the phantom bridge.

M51's spiral shape may be the result of tidal interactions with NGC 5195. Apparently the smaller companion circles the great Whirlpool in an orbit inclined 73° to M51's disk. The latest tidal event might have occurred about 56 million years ago. Ironically, that's about the time when 75 percent of living species on earth vanished in a series of cataclysmic impact episodes.

M52

The Scorpion
NGC 7654
Type: Open Cluster
Con: Cassiopeia
RA: 23^h 24^m.8
Dec: +61° 36′
Mag: 6.9
Dia: 16′
Dist: 5,100 l.y.
Disc: Messier, 1774

MESSIER: [Observed 7 September 1774] Cluster of very faint stars, mingled with nebulosity, which may be seen only with an achromatic refractor. It was while observing the comet that appeared in that year that M. Messier saw this cluster, which was close to the comet on 7 September 1774. It is below the star d [Flamsteed 4] in Cassiopeia. Star d was used to determine the positions of the star cluster and the comet.

NGC: Cluster, large, rich, much compressed in the middle, round, stars from 9th to 13th magnitude.

M52 is one of my favorite open clusters for binoculars – not because it trumpets starlight, but because it doesn't. The 7th-magnitude cluster is merely a large uniform glow; subconsciously, I must find that simplicity soothing. Although I tried very hard, I could not convince myself of seeing it with the unaided eye. The cluster lies about ½° south of 4 Cassiopeiae, a reddish 5th-magnitude star about 6° northwest of Beta (β) Cassiopeiae.

The binocular view belies the fact that M52 is a very rich telescopic

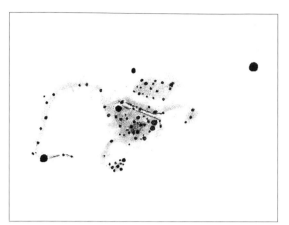

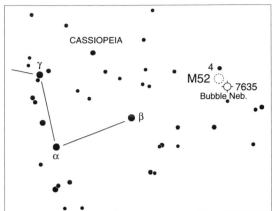

cluster. About 20 million years young, it measures 24 light years in diameter and contains nearly 200 members brighter than 15th magnitude. The cluster's computed central density is 56 stars per cubic parsec; in Messier's catalogue, only M11 and M67 are denser open clusters.

You will be immediately struck by the forceful presence of an 8th-magnitude topaz field star on the cluster's southwestern edge. This star all but leaps out at you, as if trying to steal the show. It is an imposter, however, not an actual member of the cluster. Interestingly, when I first looked at the star, and then at the cluster, which has an overall bluish hue, the topaz color of the star seemed even more pronounced. Webb and Smyth saw this star as orange, and Mallas recorded it as a conspicuous reddish star. Webb almost sounds sarcastic in his description of M52: "Irregular, with orange star, as is frequently the case," but he is merely noting that usually the most outstanding member of an open cluster shines with a ruddy hue.

With a quick glance at low power, M52 looks like a tight ball of tiny crystal chips reflecting blue light. A longer view will reveal a little isolated patch of starlight just to the northwest of a heart-shaped central body; a much larger patch lies to the southeast. With 72×, the shape of the cluster's stars looks rather arachnoid, like a scorpion. Thin wisps of faint stars jut from the body like tiny legs, and patches of starlight form the claws. The scorpion's swooping tail curves to the north, where it joins the topaz star. Two stinger stars follow to the east.

The main, heart-shaped body of starlight contains many uniformly bright stars caging a haze of fainter members, which requires peripheral vision to resolve. When you don't stare directly at this cage, do you see the thin, dark rift running along the southeastern side of the scorpion's body?

While in the vicinity, try to glimpse the elusive Bubble Nebula (NGC

7635). Only 35′ southwest of M52, this very dim ring of gas, whose brightest section surrounds an 8.5-magnitude star, is only weakly visible from dark skies. Although it resembles a planetary nebula, the Bubble is a fairly ordinary H II region, a vast cloud of ionized hydrogen.

M53

NGC 5024
Type: Globular Cluster
Con: Coma Berenices
RA: 13^h 12^m.9
Dec: +18° 10′
Mag: 7.7
Dia: 13′
Dist: 56,000 l.y.
Disc: Johan Elert Bode, 1775

MESSIER: [Observed 26 February 1777] Nebula without stars discovered below and close to Coma Berenices, close to the star Flamsteed 42 in that constellation. This nebula is circular and conspicuous. The comet of 1779 was directly compared with this nebula, and M. Messier plotted it on the chart for that comet, which will be published in the Academy volume of 1779. Observed again 13 April 1781. It resembles the nebula that lies below Lepus.

NGC: Remarkable globular cluster, bright, very compressed, irregularly round. Very much brighter toward the middle; contains 12th-magnitude stars.

William Herschel described M53 as one of the most beautiful sights in the heavens. Still, this whopping 200-light-year-wide sphere of stars is an unsung hero among globulars for small telescopes. Perhaps that is because it sits on the fringe of the Coma–Virgo Cluster of galaxies, like a flower outside a forest of redwoods. But it is a bold glow, and even binoculars will reveal its 7.5-magnitude, 13′ disk only about 1° northeast of the binary Alpha (α) Comae Berenices.

And what a haunting field M53 is in. Whenever I peer at the globular, my averted gaze continually picks up a multitude of dim spectral shapes, which seem to flitter across the low-power field. While most are close doubles, one pale ashen sheen belongs to the loose globular NGC 5053, which looms 1° southeast of M53 like the departed soul of its more brilliant neighbor.

With a glance at low power, M53 appears as a tight, unresolved ball of light with a faint outer halo. Use averted vision, however, and the inner halo, which surrounds a crisp starlike nucleus with a tangerine hue, begins to look boxy. At 72×, the inner region reveals its beauty: look for a pale gemstone inside a tiny jeweled chest, which is encased inside a larger jeweled chest; the chests are aligned east-west. High power shatters the jewel, whose opalescent fragments seem to sparkle through a thinning cloud of dust. Thus, no longer does the core look so tight, though I still see a tiny "nucleus" flanked by a wall of stars to the south. Use keen averted vision and relax your gaze. Do you see the dark rift abutting the south wall (refer to the drawing)?

Now return to 72×. The cluster appears twisted, as if someone tried to wring out the stars in the northern half. This strange sight is created by four claws of stars ripping through the outer hood to the north and west; the three northern extensions appear to be topped by a faint arc of starlight emanating from the westernmost finger. Meanwhile, the southern half of the cluster looks triangular, with the apex of the triangle pointing due west. I wonder if this is what Luginbuhl and Skiff were referring to when they wrote that "in poor seeing the core of the cluster looks roughly triangular."

Although M53 can easily be seen with binoculars, I could not positively identify it with the naked eye. I do believe someone with eyes younger than mine could. The problem is separating it from 4th-magnitude Alpha Comae Berenices.

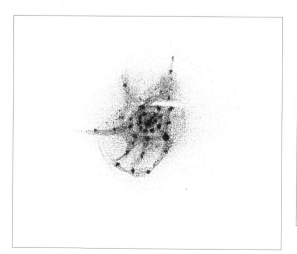

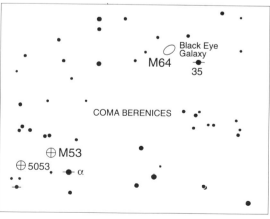

M54

NGC 6715
Type: Globular Cluster
Con: Sagittarius
RA: 18^h 55^m.1
Dec: −30° 28′
Mag: 7.7; 7.2 (O'Meara)
Dia: 12′
Dist: 70,000 l.y.
Disc: Messier, 1778

MESSIER: [Observed 24 July 1778] Very faint nebula discovered in Sagittarius. The center is bright, and it does not contain any stars; seen with an achromatic three and a half-foot refractor. Its position was determined relative to third-magnitude ζ Sagittarii.

In 1994 astronomers discovered a large, faint dwarf galaxy in Sagittarius that was superposed on the globular cluster M54. Actually, it appears that M54 belongs to the Sagittarius dwarf, which our Milky Way Galaxy is now cannibalizing. Although M54 probably formed at the same time as the dwarf galaxy, only now to be accreted into the Milky Way, some astronomers now believe that globulars are the *products* of galactic cannibalization. When two galaxies collide or merge, the interaction can either trigger star formation or leave remnant clumps of accreted galactic matter, which turn into globular clusters. Try imagining this the next time you peer at seemingly insignificant M54!

Beaming from a distance of 70,000 light years, M54 is the farthest of all the Messier globulars. Yet, it is no problem for 7×35 binoculars, appearing as a bright starlike object 1½° west-southwest of Zeta (ζ) Sagittarii, at the base of the teapot. But being so distant, the cluster does not easily reveal its secrets to telescopic viewers. In the Genesis at 23×, the globular is a softly glowing fuzzy star, sort of what Omega Centauri looks like to the naked eye. (Shapley considered M54 perfectly round.) At 72×, the 244-light-year-wide globular remains largely unresolved, though some outliers in the halo pop into and out of view – a perception shared by Skiff using a 10-inch telescope. This view confirms John Herschel's impression that the brightest stars in the cluster hover around 14th magnitude, though photometry of the cluster shows the brightest members are actually about 15th magnitude. Regardless, M54 displays a bright, starlike core surrounded by a tight yellowish inner halo that gradually diffuses outward. With persistent averted vision, the cluster looks more like a

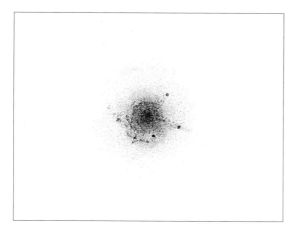

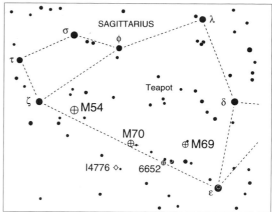

planetary nebula with a brilliant central star (something Jones noted too) than a globular. Helping to reinforce this impression is that the outer halo appears distinctly blue, glacial blue, as do some planetaries.

The view changes once again at 130×. Now M54 begins to look like a face-on spiral galaxy just beginning to show resolution. Namely, two arms of very faint stars in the cluster's northern envelope appear to swirl counterclockwise around the nucleus. Clear gaps separate these features from one another and from the core. I did note an orange star superposed on one of these arms. Can you determine which one?

NGC: Globular cluster, very bright, large, and round. Its brightness increases inward gradually, then suddenly toward the middle. Well resolved, with 15th-magnitude stars and fainter.

M55

NGC 6809
Type: Globular Cluster
Con: Sagittarius
RA: 19ʰ 40ᵐ.0
Dec: − 30° 57′
Mag: 6.3
Dia: 19′
Dist: 17,000 l.y.
Disc: Nicolas-Louis de
Lacaille, 1752

MESSIER: [Observed 24 July 1778] Nebula that appears as a whitish patch, about 6 minutes across. Its light is evenly distributed and has not been found to contain any star. Its position was determined relative to ζ Sagittarii by using an intermediate, seventh-magnitude star. This nebula was discovered by M. l'abbé de la Caille, *Mémoires de l'Académie 1755,* page 194. M. Messier searched for it in vain on 29 July 1764, as reported in his paper.

NGC: Globular, pretty bright, large, round, very rich, very gradually brighter in the middle, stars from 12th to 15th magnitude.

One of the most southerly globulars observed by Messier, M55 was included in a study of globular cluster motion, which revealed that these stellar agglomerations are moving at high speeds (greater than 100 miles per second) in elliptical orbits through the distant halo of our galaxy – like long-period comets in orbit around our sun. At a distance of only 17,000 light years, however, M55 is relatively nearby (consider that M54 is 70,000 light years distant). You can see it with the naked eye on the outskirts of the Milky Way band about 8° east-southeast of Zeta (ζ) Sagittarii. In 7x35 binoculars M55 looks like a hairy star. With the 4-inch at 23×, the cluster starts to splinter across its large and loose surface – a refreshing sight after looking at the smaller, more difficult globulars M54, M69, and M70 nearby. In fact, M55's disk looks huge when compared to the disks of those globulars, even at high power.

One night I had the pleasure of watching an earth-orbiting satellite sail across M55's misty bay of light. Such transits are becoming increasingly common to deep-sky observers. Such an event puts the scale of the universe, or our little corner of it anyway, in perspective. Here was a man-made craft about the size of a Volkswagen van and a few hundred miles away drifting in front of a cluster of stars a mind-boggling 93 light years wide and 17,000 light years distant. In my mind's eye the cluster swelled to magnificent proportions, as the moon does when it looms on the horizon.

At 72× a multitude of bright suns are scattered across a homogeneous background of fainter stars. Many of the cluster's members are nicely resolved into stellar arcs, somewhat reminiscent of the parallel streams in M25. Patterns visible in the background haze are out of sync with those formed by the brighter suns. This, combined with the way several dark lanes infiltrate the cluster from all sides, leads to a most impressive, albeit complicated, view. Ripples of stars and dark waves seem to be flowing into the cluster from the southeast, forming a cove, or

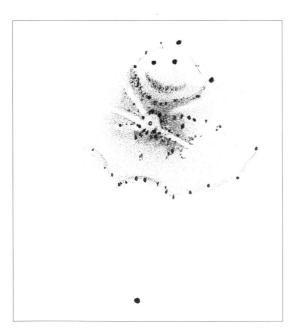

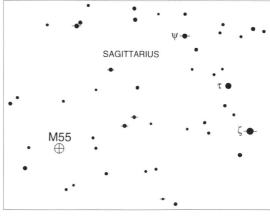

what Luginbuhl and Skiff referred to as a "large bite" out of the cluster's outer halo. Another such bite appears to have been taken out of the northwestern part of the outer halo.

The central, southern curve of stars becomes very pronounced at 130×, but I still prefer the moderate power for an overall view. That's when a bright star in a dark hole at the center of the cluster shows up the best. A strong dark bar extends equidistant southwest and northeast of that central star and dark hole. Also, dark forks branch off from the southwest axis, forming the negative image of a "peace" symbol. One doesn't expect to find a hole in the center of a compressed globular. Indeed, this is an illusion. As Burnham explains, "The unusual 'openness' of this cluster . . . is due to the fact that only a relatively small percent of the members exceed a brightness of 13th–14th magnitude, and the cluster does not begin to 'fill in' until one reaches about 17th magnitude where a vast swarm of stars quite suddenly appears."

M56

NGC 6779
Type: Globular Cluster
Con: Lyra
RA: 19^h 16^m.6
Dec: +30° 11'
Mag: 8.4
Dia: 7'
Dist: 31,000 l.y.
Disc: Messier, 1779

MESSIER: [Observed 23 January 1779] Nebula without a star, which is very faint. M. Messier discovered it on precisely the same day as he discovered the comet of 1779, on 19 January. On the 23rd, he determined its position by comparing it with Flamsteed 2 Cygni. It is close to the Milky Way and near it is a tenth-magnitude star. M. Messier plotted it on the chart for the comet of 1779.

NGC: Globular cluster, bright, large, irregularly round, gradually very much compressed toward the middle, well resolved, stars of 11th to 14th magnitudes.

With its softly glowing core, delicate spherical halo, and lack of resolution at low power, M56 has always been my favorite noncomet. I still remember many a warm summer night when I would intentionally sweep my telescope back and forth, up and down, between Beta (β) Cygni and Gamma (γ) Lyrae, until I accidentally encountered M56 about halfway between them. A chill would race up my spine whenever its cometlike form entered the eyepiece. In those magical moments I was Messier, experiencing the thrill of discovery.

Reality would nevertheless soon set in. But what a perfect snowball M56 is, 7' across and shining at 8th magnitude, amid the blizzard of the Milky Way. Actually the globular looks more like a dirty snowball, because of its grayish pallor. I find the color startling, considering that so many

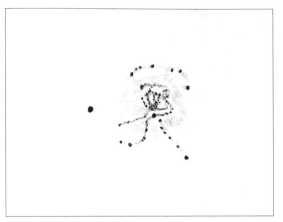

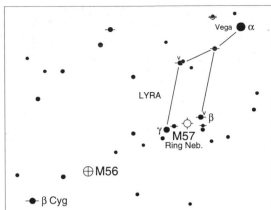

other Messier globulars display delicate pastel hues. Interstellar dust dims the cluster by about 0.2 magnitude.

In the low-power view a faint stream of stars flows from the southeast and drains into M56's foggy pool. The tiny, round glow (which is really 63 light years across) is punctuated by a 10th-magnitude star to its west. And a 5.8-magnitude, orange *M2* star about ½° to the northeast adds some color. There's also a fine 8th-magnitude double about the same distance to the southwest.

Moderate power starts to resolve the cluster, which immediately appears elongated north–south. In fact, the inner core is highly asymmetrical. With a thin fan of material to the south and nothing like it immediately opposite to the north, the innermost region of M56 is quite unique, looking like a light bulb or an exclamation mark. My pencil sketch shows the location of a rather strong, 10th-magnitude central condensation, the point of the exclamation mark. Interestingly, Mallas noted that unlike most other globulars, M56 does not have a bright core. And his drawing of the cluster shows a substantially large elliptical vacancy in the center. Jones reported that M56 has "no very marked central condensation," as did Burnham (almost verbatim). Skiff speaks of an irregularly round and broad concentration to the center, which corroborates what Smyth drew in his *Cycle of Celestial Objects*.

High power did reveal a strong dark lane separating my 11th-magnitude "central" star and the southern fan of mottled starlight. I believe this dark feature almost certainly is what Mallas observed. Still, someone should keep an eye out for a possible variable star lurking at the heart of M56. Meanwhile, don't miss out on enjoying the vast web of darkness surrounding M56 at 23×. It causes many neighboring regions to look as if the stars have been deposited there by the ebb and flow of some unseen galactic surf.

M57

Ring Nebula
NGC 6720
Type: Planetary Nebula
Con: Lyra
RA: 18^h 53^m.6
Dec: +33° 02′
Mag: 8.8
Dia: 76″
Dist: 1,140 l.y.
Disc: Antoine Darquier de Pellepoix, 1779

MESSIER: [Observed 31 January 1779] Patch of light between the stars (and (Lyrae, discovered while observing the comet of 1779, which passed very close. It seemed that this patch of light, which has rounded borders, must be composed of very faint stars. It has not, however, been possible to see them, even with the best telescopes, but the suspicion remains that there are some. M. Messier plotted this patch of light on the chart for the comet of 1779. M. Darquier, at Toulouse, discovered this nebula when observing the same comet, and he reported "Nebula between (and (Lyrae, it is extremely faint, but perfectly outlined. It is as large as Jupiter and resembles a fading planet."

NGC: A magnificent object. Annular nebula, bright, pretty large, considerably extended, in Lyra.

> Among the curiosities of the heavens should be placed a nebula that has a regular, concentric, dark spot in the middle.
>
> – William Herschel

When a star with a mass similar to that of our sun nears the end of its life, it blows off a shell of gas that, from our perspective, appears like a ring centered around the dying star. M57, the Ring Nebula, represents the remains of one such disgorging episode about 20,000 years ago. The first planetary nebula discovered, it has worked its way ever since into the hearts of virtually all telescopic observers. And rightly so, because no other planetary appears so distinctive in small apertures.

It is a challenging binocular object, well placed in the northern sky about 6½° southeast of brilliant Vega (Alpha [α] Lyrae), and nearly halfway between the eclipsing binary star Beta (β) Lyrae (whose brightness fluctuates between magnitude 3.3 and 4.3 every 12.9 days) and 3rd-magnitude Gamma (γ) Lyrae. Telescopically, M57's tiny 9th-magnitude annulus of gray smoke floats against a rich Milky Way field crisscrossed with dark streamers, some of which appear to be as gray and smoky as the Ring Nebula itself.

The "ring" is actually a torus (doughnut-shaped) viewed looking down the hole. This is unlike the planetary M27, which is seen side on. The 0.4-light-year-wide gaseous wreath of M57 was likely blown off the white-dwarf central star some 20,000 years ago and is still expanding at a rate of about 16.5 miles per second, or about 1″ per century.

One night when I viewed M57 nearly overhead at 72×, I found the sight almost shocking. Not only was the "hole" clearly visible but the entire ring appeared to scintillate with stellar jewels – foreground stars superposed on the gaseous loop. As my eye raced to see one star in the nebula, my averted vision captured another. As soon as that one was spied, my averted vision jumped to the next one. In this way the stars seemed to *revolve* around the ring like the flashing of light bulbs around a Las Vegas billboard. Some nineteenth-century observers alluded to seeing this effect. For example, astronomer Angelo Secchi believed he resolved the ring into minute stars "glittering like stardust," and Webb saw M57's light as "unsteady . . . probably an illusion."

The brightest star in the gaseous ring shines at 13th magnitude and is located just 1′east of center, 20″ beyond the extreme outer rim. A 14th-magnitude star lies just inside the Ring's southwest rim and is commonly mistaken for the central star. In most star catalogues the blue-dwarf central star is listed as 15th magnitude, but its brightness is the subject of considerable debate. About 80 years of photometric data reported in the *Catalogue of the Central Stars of True and Possible Planetary Nebulae* (Acker and Gliezas, Observatory of Strasbourg, 1982) show the central star varying from magnitude 13 to magnitude 15. Skiff notes that several recent photoelectric measurements give values of magnitude 15.0 to 15.2. Many veteran skywatchers have estimated it to be as bright as 14th magnitude (putting it within range of a good 4-inch telescope under a dark sky). Yet Burnham notes that the star was fainter than 16th magnitude when he looked at it in 1959 through the 40-inch reflector at Lowell Observatory. Although the star appears to be variable, this may only partly explain the

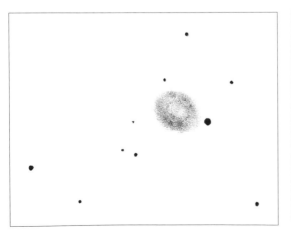

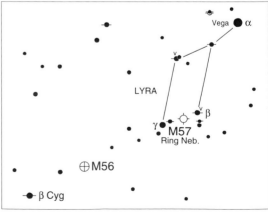

disparate magnitude estimates. One summer evening in 1988, my wife, Donna, and I were visiting comet discoverer Michael Rudenko at the 18-inch refractor at Amherst College in Massachusetts. Some observers had turned the telescope to the Ring Nebula and were complaining that they could not see the central star. When I looked through the eyepiece, I noticed that the apparent size of the Ring was so small that the "hole" actually appeared bright, diminishing the contrast between the star and its background. I removed that eyepiece and replaced it with another one that provided about 1,000× magnification. And – voilà! – the central star burst into view, suggesting that the apparent brightness of the star *is* affected by the surrounding nebulosity.

I also looked at the Ring through two 40-inch reflectors, one atop Pic du Midi in the French Pyrenees, the other at Lick Observatory on Mount Hamilton, California. At Pic du Midi, William Sheehan and I centered M57 in the finder scope, but were then surprised that we could not see it in the main telescope; all we saw was a pair of stars. It turned out that at 1,200× the gaseous ring was outside our tiny field of view; we were seeing only the central star and its similarly bright neighbor to the northwest, which allowed us to conclude that the central star shines at around magnitude 14.5. Interestingly, though I saw the central star through the 40-inch reflector at Lick Observatory, I did so only with great difficulty – the view was at low power, and I was not in a position to change the eyepiece. Once again, the contrast between the star and its background was diminished. Now, consider that on 2 September 1981, Archinal made a detailed drawing of M57 based on his observations with Lowell Observatory's 72-inch reflector on Anderson Mesa in Arizona, which shows that the central star's neighbor was noticeably fainter than the central star (or the central star was noticeably brighter than the neighboring star), perhaps by 1 to 2 magnitudes! He concludes that the central star is between 15th and 16th magnitude. "When I've had problems making it out," he reports, "I can tell that it's usually due to poor seeing or poor-quality large-aperture telescopes or eyepieces."

Houston argued that the ability to see the central star is also enhanced by one's color sensitivity: those with blue-sensitive eyes are more apt to see it. He confirmed this for himself after having a cataract operation, where his yellowed lens was replaced with a fresh transparent one, thus allowing more ultraviolet radiation into the eye. I have seen the central star with the 9- and 15-inch refractors at Harvard College Observatory at high magnifications. And I'm convinced a skilled observer could see it in even smaller telescopes – ones that can tolerate high magnifications well. A steady atmosphere is also required.

But high power really is necessary because, as John Herschel noticed, the interior of the ring is "filled with a feeble but very evident light." And Rosse, through his great reflector, saw striations. The trick to perceiving Herschel's feeble light with a small aperture is to use 23× and 72× and compare the intensity of the interior of the ring with that of the sky just outside the ring. Some observers stop "looking" after they see the Ring's dark hole, missing the opportunity to see the weak interior glow.

Finally, I always like to end an observing session by returning to low power and "relaxing." When I do this with M57, I see the ring as an inflated inner tube afloat in a semicircular black pond. Now concentrate on the dark "water." Do you see glints of moonlight reflecting off the wave crests?

M58

NGC 4579
Type: Barred Spiral Galaxy
Con: Virgo
RA: $12^h 37^m.7$
Dec: $+11° 49'$
Mag: 9.6
SB: 13.0
Dim: $5'.9 \times 4'.7$
Dist: 55 million l.y.
Disc: Messier, 1779

MESSIER: [Observed 15 April 1779] Very faint nebula discovered in Virgo, almost on the same parallel as third-magnitude (. The slightest illumination of the micrometer crosshairs causes it to disappear. M. Messier plotted it on the chart for the comet of 1779, which will be found in the Academy volume for the same year.

NGC: Bright, large, irregularly round, very much brighter in the middle, mottled.

Modestly bright and unassuming, M58 is, nonetheless, among the most prominent Messier galaxies in the Virgo Cloud. Spanning 94,000 light years, this barred spiral has about the same mass as our Milky Way Galaxy, or about half that of the Andromeda Galaxy (M31). You will find it stationed 8′ east of an 8th-magnitude star, about 2° northwest of 4.9-magnitude Rho (ρ) Virginis. If you live under dark skies try locating M58 first with binoculars; look for a mere glint of diffuse light. The galaxy shares the same low-power telescope field with the bright ellipticals M59 and M60; it

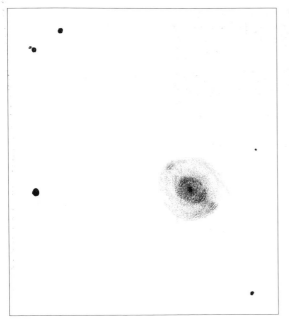

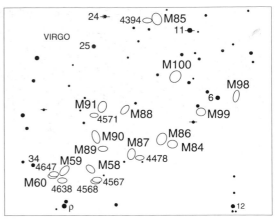

is also north of a most intriguing double galaxy, NGC 4567 and NGC 4568, also known as the Siamese Twins. At 23×, M58 displays only a starlike nucleus surrounded by a uniform halo that gradually fades into the background. But 72× immediately brings out a mottled halo and a patchy inner glow, and 130× starts to reveal some of the galaxy's subtle structure.

When looking at galaxies with high‑ power, I usually alternate between using hyperventilation (see page 23) and relaxation techniques to visually chisel out detail. For M58 the first features to materialize out of the galactic haze are knots on the east and west ends of the nuclear region, followed by the dim extensions that outline the galaxy's gently curving arms. Only the northeastern part of the bar connecting the outer knots to the nucleus is definitely visible in the 4-inch; the bar to the southwest was merely a suggestion, probably because our eyes like to create patterns. Jones, however, reported clearly detecting the bar from England in an 8-inch telescope!

Concentrate on the innermost nuclear region. Here I do see a bar connected to faint arcs. The details in this inner region wonderfully replicate those in the outer halo, only they're just slightly skewed.

M59 and M60

M59

NGC 4621
Type: Elliptical Galaxy
Con: Virgo
RA: 12^h 42^m.0
Dec: 11° 39′
Mag: 9.6
SB: 12.5
Dim: 5′.4 × 3′.7
Dist: 55 million l.y.
Disc: Johann Gottfried Kohler, 1779

MESSIER: [Observed 15 April 1779] Nebula in Virgo and close to the preceding one [M58], on the parallel of (, which was used to determine its position. It is of similar luminosity to the previous nebula, being just as faint. M. Messier plotted it on the chart for the comet of 1779.

NGC: Bright, pretty large, little extended, very suddenly very much brighter in the middle, two stars preceding [westward].

The two elliptical galaxies M59 and M60 fit neatly in the same field of view with M58 in 7×35 binoculars. M60 is clearly the brightest of this trio in Virgo. But this revelation caused me alarm, because Jones had recorded M60 almost a full magnitude *fainter* than M58 (8.8 and 9.6, respectively), and Mallas had found the two to be of similar magnitude (9th). Fortunately, Skiff came to my rescue, noting that the *Second Reference Catalog of Bright Galaxies* (Gerard de Vaucouleurs and A. de Vaucouleurs, University of Texas Press, 1976) lists M60 as being nearly a full magnitude brighter than M58, which agrees with my estimate.

I cannot help but think of an elliptical galaxy as little more than a uniformly bright and structureless glow. But my impression is one created largely by photographic images. So I am always surprised (and somewhat incredulous) whenever I look at an elliptical galaxy through a telescope and actually see detail. These statements probably say a lot about the power of photography to alter our visual perception.

For example, I was surprised to see that M59 contains a sharp star-like nucleus, because you hardly ever see one in a long-exposure photograph. (I wasn't aware that John Herschel had noted M59 was "very suddenly much brighter in the middle," and that Jones had also detected the starlike core.) M59 also looks mottled, though I could not tell whether this is because of faint stars superposed on it, or from actual dust lanes and patches, or from physiological effects. The brightness across the disk is definitely asymmetrical, with the northwest quadrant decidedly brighter than the southeast. Most puzzling is what appears to be a faint, needlelike bar running through its major axis – a feature I have seen associated with other ellipticals. It is probably just an illusion, though. Is there any way to verify the existence of such a perplexing sight?

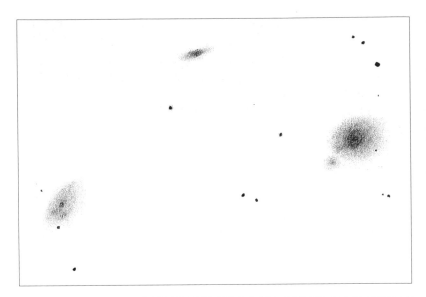

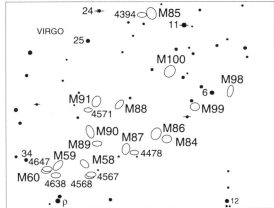

Among the largest ellipticals, M60 measures 118,000 light years across and is some 200 times more massive than the elliptical galaxy M32, the more condensed companion of the Andromeda Galaxy. By comparison, M59 is 32,000 light years smaller than M60. Like M59, M60 also has a starlike nucleus. Admittedly, when I first saw M60's crisp central star, then thought of its fuzzy photographic image, and then noticed that the galaxy was brighter than 9.6-magnitude M58, how could I not suspect a supernova? But, that isn't the case. As Luginbuhl and Skiff note, 8.8-magnitude M60 is an evenly fading circular glow with a faint central pip. M60 has a tiny companion, NGC 4647, 2′.5 to the northwest, whose seemingly transparent glow is a pleasing sight. This galaxy duet reminds me of Alcor and Mizar, the famous naked-eye optical double star in the handle of the Big

Dipper. Unlike M60, NGC 4647 is a face-on spiral comprising 10 billion suns. Webb missed it with a 3.7-inch telescope, which, I am certain, says more about the hazy England skies than about the observer himself.

But consider the following experience. Midway between M59 and M60, and to the south, is the 12th-magnitude elliptical galaxy NGC 4638. I find that by staring directly at NGC 4638 at 130×, M60's tiny companion burns into view in my peripheral vision! But, when I look directly at M60, the companion disappears. Thus I refer to NGC 4647 as the "disappearing galaxy." With averted vision NGC 4647 really swells and displays definite inner and outer sections. It also looks gray.

Photometric data show the colors of M59 and M60 to be almost identical, but to me M59 looks bluer than M60, which has a yellower tint. In fact, if you use high power and relax, M60 appears to have an intense yellow core that seems to be surrounded by countless glistening beads forming a diamond-shaped structure around it.

M60

NGC 4649
Type: Elliptical Galaxy
Con: Virgo
RA: 12^h 43^m.7
Dec: + 11° 33′
Mag: 8.8
SB: 12.8
Dim: 7′.4 × 6′.0
Dist: 55 million l.y.
Disc: Johann Gottfried Kohler, 1779

MESSIER: [Observed 15 April 1779] Nebula in Virgo, slightly more conspicuous than the previous two [M58 and M59], again on the same parallel as (, which was used to determine its position. M. Messier plotted it on the chart for the comet of 1779. He discovered these three nebulae when observing the comet, which passed very close to them. The latter passed so close on 13 and 14 April that both were in the same field of the telescope, but he was unable to see the nebula. It was not until the 15th, when searching for the comet, that he perceived the nebula. These three nebulae do not appear to contain any stars.

NGC: Very bright, pretty large, round, the following [eastern] member of a double nebula.

M61

Swelling Spiral
NGC 4303
Type: Spiral Galaxy
Con: Virgo
RA: 12ʰ 21ᵐ.9
Dec: +4°28′
Mag: 9.6
SB: 13.4
Dim: 6′.5×5′.8
Dist: 55 million l.y.
Disc: Barnabus Oriani, 1779

MESSIER: [Observed 11 May 1779] A nebula that is very faint and difficult to see. M. Messier [mis]took this nebula for the comet of 1779 on 5, 6, and 11 May. On the 11th he realized that it was not the comet, but a nebula that happens to lie on its path and at the same point in the sky.

NGC: Very bright, very large, very suddenly brighter toward the starlike center, binuclear.

Here we have one of the most pleasing open-faced spirals in the Messier catalogue for small telescope users. This may be due to its small apparent size – 6′.5×5′.8 – which condenses the galaxy's light, making the features more apparent. With a true diameter of about 100,000 light years, though, M61 is one of the larger galaxies in the Virgo Cloud. A magnificently detailed system with crooked arms, it is easily located 5° north and slightly east of Eta (η) Virginis, halfway between the 6th-magnitude stars 16 and 17 Virginis. Although I could detect its feeble 10th-magnitude light with 7×35 binoculars, it is essentially at the binocular limit, and I imagine would be difficult to glimpse in skies that were not very clear and dark.

The *NGC* description uncharacteristically misled me into thinking M61 would appear brighter and larger than it really is. At 23× the galaxy is so small and faint that I almost swept over it, perhaps because it is tucked away in the crook of a nice wishbone asterism of 9th-magnitude stars. Take the time to concentrate on M61's light, which with averted vision appears to gradually dilate into an appreciable disk. Don't be surprised if your eye also begins to notice other ghostly glows hiding among a forest of similarly bright stars populating the region. These are the dim and distant shapes of at least a dozen nearby NGC galaxies!

M61 itself reveals a brightening at the center that pops into view only with averted vision. This nucleus is centered in a fairly condensed halo of light that is wrapped in an even fainter shell. The fading into view (or out of view, depending on how you stare at it) of this faint outer envelope is the reason the galaxy appears to swell (and shrink), which is why I refer to M61 as the Swelling Spiral.

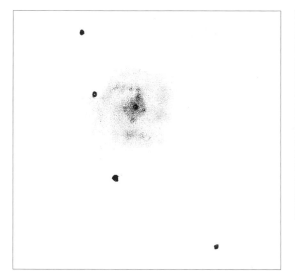

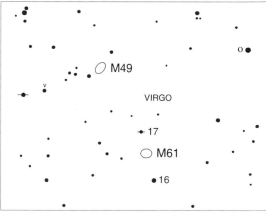

Moderate power does not add much to the view, though it does start to reveal star-forming regions, which appear as knots in the outer halo. But because this galaxy is so tight, high power is required to resolve it into its separate parts. And what a very interesting view it is at 130x! A starlike nucleus is surrounded by a mottled, diamond-shaped inner core. With averted vision the northernmost knot shows an extension that trails off irregularly to the east in a wavelike fashion, like a gracefully thin wake of smoke. Closer to the nucleus and to the south is another star-studded region that trickles off in lumps to the west before it curves sharply to the north. A faint but definite arm can be glimpsed to the east, and it boxes in the nuclear region. Overall, at high power, the galaxy looks like a square with slightly rounded edges. Through his 4-inch refractor, Mallas also observed three luminous patches outside the nucleus, which match the positions of M61's spiral arms seen in photographs.

Here's a challenge. Try to detect the three stars running along the western edge of the galaxy. The southernmost star is the most obvious at 14th magnitude. The northernmost one shines closer to 15th magnitude. The one in the middle is roughly magnitude 14.5.

M62

Flickering Globular
NGC 6266
Type: Globular Cluster
Con: Ophiuchus
RA: 17^h 01^m.2
Dec: −30°07′
Mag: 6.4; 6.7 (O'Meara)
Dia: 11′
Dist: 19,560 l.y.
Disc: Messier, 1771

MESSIER: [Observed 4 June 1779] A very fine nebula, discovered in Scorpius; it resembles a small comet. The center is bright and is surrounded by faint luminosity. Its position was determined relative to the star (Scorpii. M. Messier had seen this nebula before on 7 June 1771, but was only able to determine the approximate position. Observed again 21 March 1781.

NGC: Remarkable globular, very bright, large, gradually much brighter toward the middle, well resolved, stars of 14th to 16th magnitude.

Odd and mystifying are the words that seem to best describe M62 in Ophiuchus. A 6.7-magnitude globular 63 light years wide, it is singled out in the literature mostly for its strong asymmetry. John Herschel, Jones, and Shapley all noted its curious shape, as now I do. M62 is the closest Messier globular to the nucleus of the Milky Way Galaxy, so the strong gravitational tug on the globular might account for its peculiar visage. Specifically, the western half of the cluster is dramatically brighter than the eastern half. Although Mallas was aware of M62's asymmetry, he could not convince himself of seeing it. I don't understand why. Even at 72×, the western half stands out boldly against the stars of the Milky Way, while the cluster's weaker, eastern stars quickly fade as you move away from center. At the eastern edge, several strings of stars dangle like loose threads in a tattered cloth. Looking at it differently, that shredded eastern section makes M62 appear like a still photograph of a balloon just beginning to burst, with the "air" rushing out to the east-southeast.

But I find M62 mystifying for reasons other than its asymmetry, which I have seen in other globulars (to a lesser degree), such as M56 in

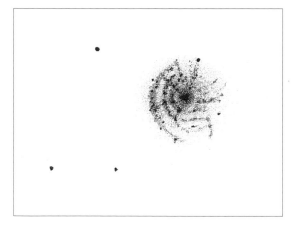

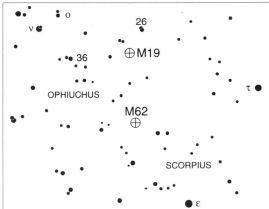

Lyra. What fascinates me is that the core of M62 appears to flicker, like a dying flame. The effect is hypnotic. One night, while watching that flame dance at 72×, I quickly broke the spell and changed to an eyepiece yielding 130×. Still, my gaze was magically drawn toward the center! The core retained its ruddy cast and continued to pulsate. Seeking an explanation for this intriguing effect, I soon figured out that it was not real – the core isn't actually pulsating in brightness – but an optical illusion.

The western section of the tight inner core has three distinct waves of starlight that seem to ripple out from the ruddy nucleus (see the drawing). Given the wealth of detail in such a small area, my eye cannot help but leap outward from wave crest to wave crest and then return to the nucleus, where the process repeats. As that is happening with my eyes, my mind is trying to fathom another illusion: faint rings of starlight appear to be rippling out from the core and *toward* the viewer! These changing optical moiré effects set the cluster's core into seeming animation.

Equally intriguing is that as the nucleus "flickers," its color changes like that of a flame in a breeze; one moment it is smoky yellow, the next it's dusky red. This, I believe, results from the eye shifting rapidly back and forth between the reddish center and the yellowish inner halo. The outer envelope also wavers with a pale bluish light. These effects do not appear to be atmospheric (the cluster is some 40° high from my observing site), but physiological (differential refraction can also cause color changes). I saw a similar visual effect one day when staring at a neighbor's needle-point of a cat.

To find this fascinating object – one of the brightest globulars in Ophiuchus – look about 5° (two finger-widths) northeast of Epsilon (ε) Scorpii, a 2nd-magnitude, orangish *K*2 star at the beginning of the Scorpion's tail. Can you see the globular with the naked eye?

M63

Sunflower Galaxy
NGC 5055
Type: Spiral Galaxy
Con: Canes Venatici
RA: 13^h 15^m.8
Dec: + 42° 02'
Mag: 8.6
SB: 13.6
Dim: 12'.6 × 7'.2
Dist: 23.5 million l.y.
Disc: Pierre Méchain, 1779

MESSIER: [Observed 14 June 1779] Nebula discovered by M. Méchain in Canes Venatici. M. Messier searched for it; it is faint and is approximately as bright as the nebula described here under number 59. It does not contain any stars and the slightest illumination of the micrometer's crosshairs causes it to disappear. Close to it there is an eigth-magnitude star, which crosses the meridian crosshair before the nebula. M. Messier plotted the position on the chart for the comet of 1779.

NGC: Very bright, large, and pretty much extended in position angle 120°. Very suddenly much brighter in the middle toward a bright nucleus.

In photographs M63 looks like a spiral galaxy that has lost control of its gravity, and we are catching a rare sight of its arms being tossed into space. This spiral, which shines with the light of 10 billion suns, is a prime example of a type of galaxy that displays a lack of cohesion between its inner and outer arms. The inner region of M63's 86,000-light-year-wide disk is ringed by strong spiral structure, while the plentiful outer arms appear loose, patchy, and haphazard. Not surprisingly, M63 is nicknamed the Sunflower Galaxy because of its resemblance to that towering plant whose dense, seedy head is ringed by an abundance of bright, overlapping petals.

To find this special-case Messier galaxy, use low power and sweep about 2° north of 20 Canum Venaticorum, a 4.7-magnitude star about 5½° northeast of Alpha^{1,2} ($\alpha^{1,2}$) Canum Venaticorum, a fine double composed of a blue 2.9-magnitude primary and a white 5.5-magnitude secondary separated by 19". At first glance, the galaxy might appear shy, as if it is trying to hide behind the fiery blaze of an 9th-magnitude star just 4' to the west. Take advantage of this illusion to "see" this galaxy beyond the stars of our own system.

Through a telescope at 23×, its soft glow reveals hints of a mottled structure in the pale outer disk, while the core is tack sharp. The challenge is to use moderate and high power to make sense of the arms. This will probably require several nights at the telescope making drawings. The problem I encountered when starting my drawings was that I could not instantly determine which way the spiral arms wind around the core – clockwise or counterclockwise – all the details are so delicate. In fact, M63

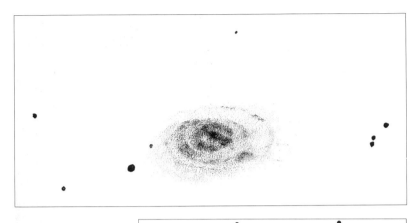

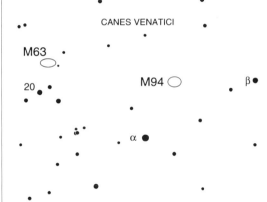

is one of the most finely detailed galaxies in the Messier catalogue in small telescopes. You will need patience. I start by focusing in on the innermost glow, where the spiral pattern is more strongly suggested. It almost helps to blur your vision or defocus the image in the telescope ever so slightly, so that the patches blend into arms. Archinal notes that this resolution problem is similar to that observed with M51, where early observers noticed only a *ring,* not spiral arms.

With 130×, I noticed a strange alignment of patches near the nucleus, which I call the "crooked cross." First, I saw four prominent condensations in a dappled ring surrounding the starlike nucleus. When I concentrated on the core, suddenly a bar seemed to cross the nucleus from southwest to northeast. This bar appears skewed about 30° from the galaxy's major axis and lines up with two of the bright knots just described. The other two bright condensations, however, do not quite line up to form a straight cross. No matter, it is all an illusion, but try and see it, because it is a good example of how Percival Lowell was able to "see" canals on Mars – his eye and brain played a game of connect the dots.

The Messier Objects 189

M64

Black Eye Galaxy
NGC 4826
Type: Spiral Galaxy
Con: Coma Berenices
RA: 12^h 56^m.7
Dec: +21° 41′
Mag: 8.5
SB: 12.4
Dim: 10′ × 5′.4
Dist: 13.5 million l.y.
Disc: Johan Elert Bode, 1779

MESSIER: [Observed 1 March 1780] Nebula discovered in Coma Berenices, which is slightly less apparent than the one that is below the hair [M53]. M. Messier plotted the position on the chart for the comet of 1779. Observed again 17 March 1781.

NGC: Remarkable, very bright and large, greatly elongated in about position angle 120°. Has a brighter middle with a small bright nucleus.

In photographs M64, the famous Black Eye Galaxy in Coma Berenices, is a very distinctive spiral system. Its smooth, silken arms wrap gracefully around a porcelain core, whose northern rim is lined with dust. The galaxy resembles a closed human eye with a "shiner." The dark dust cloud looks as thick and dirty as tilled soil. But in his classic book, *Galaxies*, astronomer Timothy Ferris notes that a jar of its material would be difficult to distinguish from a perfect vacuum. Yet M64's black cloud is so expansive – some 40,000 light years in diameter – that it contains enough atoms to loam the gardens of billions of planets. Appreciate the soil you walk on, Ferris reminds us, because every atom in it once belonged to an interstellar dust cloud like the one we see so strikingly in M64.

Visible as a gentle glow in 7 × 35 binoculars, M64 is an easy catch less than 1° east-northeast of 35 Comae Berenices. Through a telescope this 8th-magnitude spiral appears as elegant as it does in photographs, though not as detailed. Its brilliant nucleus lies within an extremely smooth oval disk with a milky texture. The disk also has a hint of blue coloration. I found the view at 72× very confusing: I was expecting to see a large, uniform glow, but so much subtle detail suddenly materialized that I thought I must have chanced upon the wrong galaxy.

The debate over the visibility of the dark cloud forming the "black eye" is nearly as intense as the debate about the visibility of the central star in M57. Here are the extremes: Jones felt that an 8-inch telescope is needed to see it with certainty, whereas Mallas claimed to have spied it in a 2.4-inch glass! The first time I looked at M64 with the 4-inch Genesis, I resolved the black eye without realizing it. That is, I noticed the feature

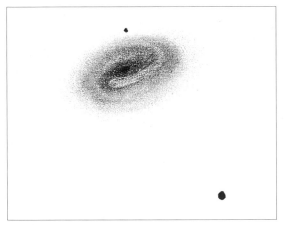

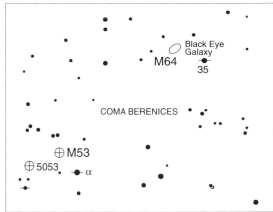

only upon studying my *drawing:* first I noted an area in the drawing lacking detail, then I realized that the shape of this void matched the location of the black eye! Therefore, I recorded it indirectly. The feature certainly is not visually obvious in small telescopes, so don't let the photograph fool you. Use high power to search for the black eye, because the apparent size of the galaxy ($10' \times 5'.4$), and especially its inner core, is small. You need magnification to get in there and separate the inner arms. The galaxy's outer details, though, have too low contrast for high power, and show up best at $72\times$.

I was surprised to read Luginbuhl and Skiff's comment on the galaxy's "non-stellar nucleus." My impression was quite different: I perceived a very bright stellar nucleus. Supporting Skiff's view is a drawing by Mallas in *The Messier Album* that shows nothing even resembling a core. Contrary to that, however, the *NGC* description states that M64 has a "small, bright nucleus." Jones sided with Luginbuhl and Skiff: "the central nucleus is small but decidedly not starlike." John Herschel said, "I am much mistaken if the nucleus be not a double star." This quote originates from Herschel's 1833 log notation describing M64 as "vsmbm," meaning "very suddenly much brighter toward the middle." In another observation Herschel says it is "very suddenly very much brighter toward the middle, almost to a star, but magnifying destroys this effect." These observations suggest that the appearance of M64's core depends on the magnification used.

M65

NGC 3623
Type: Spiral Galaxy
Con: Leo
RA: 11^h 18^m.9
Dec: + 13° 05'
Mag: 9.3; 8.8 (O'Meara)
SB: 12.4
Dim: 9'.8 × 2'.9
Dist: 24 million l.y.
Disc: Pierre Méchain, 1780

MESSIER: [Observed 1 March 1780] Nebula discovered in Leo. It is very faint and does not contain any stars.

NGC: Bright, very large, much extended in position angle 165°, gradually brightening to a bright central nucleus.

Leo's M65 and M66 probably rank second only to M81 and M82 as sought-after galaxy pairs. Located about halfway between Theta (θ) and Iota (ι) Leonis and separated by only 21', M65 and M66 are both visible in 7 × 35 binoculars – as is a larger, fainter, edge-on galaxy – NGC 3628 – just 35' north-northeast of M66. Through the 4-inch, all three galaxies vie for attention in the same low-power field. These three galaxies may be part of an independent cluster of galaxies on the near edge of the vast Virgo Cloud.

At 23 × M65 appears immediately oval and quite large. To me, it looks a half-magnitude brighter than M66, but some sources claim just the opposite. For example, Jones asserted that M66 is actually the brighter, but that M65 tends to appear more conspicuous because of its streaky outline. At a glance, it might appear that the discrepant brightness estimates are related to the surface brightness of the galaxies. Surface brightness is, in a very general sense, akin to dividing the object's magnitude by its area. The surface brightness of M65 and M66 is roughly the same, 12.4 and 12.5, respectively; this means that each arc minute of the galaxy shines roughly with the brightness of a 12.5-magnitude star. So the difference in surface brightness between M65 and M66, a mere 0.1 magnitude per square arc minute, is not sufficient to account for the brightness discrepancy. Without question, in photographs the tightly wound mass of M66, with its burning core and clumps of star-forming regions, looks brighter than the more normal-looking spiral M65. But photographic images do not reflect what is seen visually. I also made my magnitude estimate with 7 × 35 binoculars, so as not to be fooled by the telescopic appearances.

An optical phenomenon can make M65 appear larger than it really is

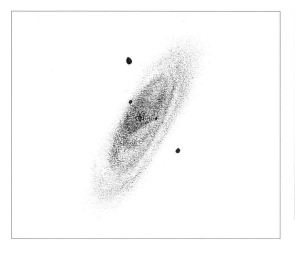

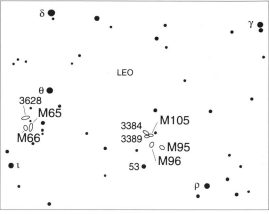

at low power. I found that at first glance the galaxy looks very long and extended – the tips of the major axis seem to trail faintly off to unknown lengths. But when I concentrate on the more brightly illuminated area, the galaxy appears quite stubby. This illusion of infinite length is caused by the dust lane that sharpens the galaxy's eastern limb, which the eye likes to extend.

M65 is a joy to see but an awfully difficult galaxy to observe. The details within its bright, nearly edge-on disk vary only slightly from the background brightness, so they're hard to pick out from the "noise." But they can be recorded with accuracy once you familiarize yourself with the galaxy after a few observing sessions. Just focus on a different part of the galaxy each night. Use 72 × to survey the diamond-shaped nuclear region, which is partly an illusion caused by two faint stars nearby – a 12th-magnitude one southwest of the core and a 13th-magnitude star to the northeast. Aside from the dust lane, the most prominent feature lies northwest of the nucleus, where the galaxy's arms can first be seen as faint patches, then in three distinct spirals. The details in the southwest section are far more difficult and less defined. Look for some knots just outside the nuclear region connected by a looping arc.

Visually, the details of M66 are even more subtle than those of M65. Like the Black Eye Galaxy (M64), M66 appears very soft and graceful; it's nothing like the strong and dynamic image you see in photographs. The galaxy's bright, starlike nucleus is its most noticeable feature. Use low power to compare the cores of M65, M66, and NGC 3628. M65 has a somewhat stellar nucleus, and NGC 3628 reveals absolutely nothing! Now look for a bright knot immediately northwest of M66's nucleus. A dark patch lies east of the knot, followed by a mere stump of a spiral arm. The region

M66

NGC 3627
Type: Spiral Galaxy
Con: Leo
RA: $11^h 20^m.2$
Dec: $+12° 59'$
Mag: 9.0
SB: 12.5
Dim: $9'.1 \times 4'.2$
Dist: 21.5 million l.y.
Disc: Pierre Méchain, 1780

MESSIER: [Observed 1 March 1780] Nebula discovered in Leo; very faint and very close to the preceding one [M65]. They both appear in the same telescopic field. The comet observed in 1773 and 1774 passed between these two nebulae on 1 and 2 November 1773. Doubtless M. Messier did not see it then because of the comet's light.

NGC: Bright, very large; much extended in position angle 150°; much brighter in the middle; two stars northwest.

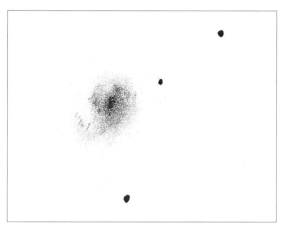

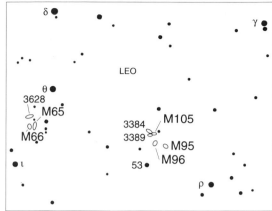

surrounding the nucleus is oval shaped and oriented southeast to north-west. See if your eye doesn't catch a streak of light running off to the south. The orientation of this streak can help you determine the galaxy's spiral pattern. The rest of the details are a chaotic mix of faint streaks of dark and light. But don't let this confusion deter you from drawing the detail, because, if you look at the photograph, the galaxy really is a dizzying world of curdled starlight.

By the way, I call NGC 3628 the "Vanishing Nebula," because with each increase of power, the galaxy blends more and more into the background of deep space, until it all but vanishes. This happens because the thick dust lane, which runs across the galaxy's entire length, overpowers the feeble light outlining it. So as you increase power, you magnify the dark lane, while spreading the faint light across a larger area of sky. At magnitude 9.5, NGC 3628 shines almost as brightly as does M65 and M66, but it has a much lower surface brightness of 13.7.

M67

King Cobra
NGC 2682
Type: Open Cluster
Con: Cancer
RA: 8ʰ 51ᵐ.4
Dec: + 11° 49′
Mag: 6.9; 6.0 (O'Meara)
Dia: 25′
Dist: 2,600 l.y.
Disc: Johann Gottfried Kohler,
between 1772 and 1779

MESSIER: [Observed 6 April 1780] Cluster of faint stars with nebulosity, below the southern claw of Cancer. Its position was determined from the star α.

NGC: Remarkable cluster, very bright and large, extremely rich, little compressed, stars from 10th to 15th magnitude.

M67 is a conspicuous naked-eye open cluster 2° west of 4th-magnitude Alpha (α) Cancri, though it is often neglected for Cancer's brighter, showier Beehive Cluster (M44), 9° to the north. By way of comparison, M67 measures 25′ across and shines at magnitude 6.0 (by my estimate), whereas M44 measures 70′ across and beams more brightly at magnitude 3.1. Try to detect these clusters simultaneously with the unaided eye, because they can give you a sense of intergalactic depth: M44 is 515 light years distant while M67 is 2,600 light years distant, or five times farther away! The diameter of M67 is 19 light years – about twice that of M44.

At somewhere between 4 and 5 billion years old, M67 is one of the most ancient open clusters known. The cluster contains more than 300 stars between 10th and 16th magnitude, and has a core density of 27 stars per cubic parsec. At 23×, the cluster first appears as a loose and uniform sprinkling of bright stars across a carpet of fainter suns. But the view quickly changes into one of a slightly oval sphere of stars separated from another stellar clump to the south-southeast. Far afield, weak arms of stars fly off to the north, but are these associated with M67 or are they just chance alignments?

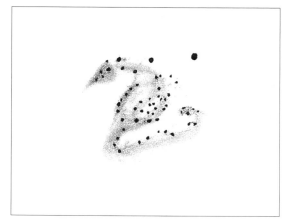

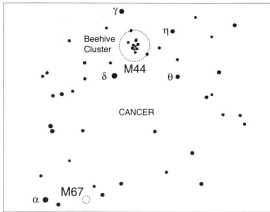

With a little imagination, the cluster looks as if it is dangling from a rack of three relatively bright field stars, the brightest member of which shines at 8th magnitude with incandescent yellow light. A dark rift runs east of that star, where it is rimmed to the north by a faint string of stars. More than the Beehive, M67 looks like a swarm of insects that seem particularly attracted to the light of the yellow star. The faintest suns cannot be resolved in a 4-inch, but many huddle together in a "nebulous" mass. As a whole, M67's stars form a distinct reverse S-shaped pattern, whose swollen (hazy) midsection reminds me of a king cobra that has just swallowed a meal. A pointed arrow of stars in the southern stellar clump forms the snake's head, and the faint stream of stars to the north marks its tail.

I hope this cluster finds its way into your mind's eye better than Smyth's "Phygian cap" asterism did to Jones's, or to mine for that matter. As Jones said, "The stars [of M67] form a pleasing pattern but Smyth's description of it being like a Phygian cap – a high conical head-dress like a bishop's mitre, or the 'Cap of Liberty' worn by the French revolutionaries – does not exactly leap to the eye." In case my snake asterism doesn't work for you, try Flammarion's "sheaf of corn," or Luginbuhl and Skiff's "fiber-optic tree," or try creating your own metaphor.

M68

NGC 4590
Type: Globular Cluster
Con: Hydra
RA: 12^h 39^m.5
Dec: −26° 45′
Mag: 7.3; 7.6 (O'Meara)
Dia: 11′
Dist: 31,300 l.y.
Disc: Pierre Méchain, 1780

MESSIER: [Observed 9 April 1780] Nebula without stars below Corvus and Hydra. It is very faint, very difficult to detect with refractors. Close to it is a sixth-magnitude star.

NGC: Globular cluster of stars, large, extremely rich, very compressed, irregularly round. Well resolved, stars of magnitude 12 and fainter.

If you have a good southern horizon, and are under dark skies, M68 is a marvelously challenging naked-eye globular. I estimated the cluster's magnitude to be 7.6, which is 0.3 magnitude fainter than the published photometric value. The cluster, another Méchain discovery, is located in Hydra just 45′northeast of a 5½-magnitude star, and about 4° south-southeast of 2.6-magnitude Beta (β) Corvi. In fact, if you stare at that 5½-magnitude star and concentrate on M68's position, the cluster might just pop out at you. With a dedicated effort, I could just make it out with the unaided eye and using averted vision, so it joins the ranks of M79 as being another of the fainter globulars visible without optical aid. Needless to say, M68 is a cinch in binoculars, unresolved yet obviously not a star.

At 23 × M68 is very compact but highly mottled, hinting at how nice it will appear at higher powers. Of this globular, Smyth wrote, "It is very pale but so mottled that a patient scrutiny leads to the inference that it has assumed a spherical figure in obedience to attractive forces." Take a moment to look around the cluster with east up. Do you see how it sits in a V-shaped asterism, a bucket, to the west? Now look to the east; a chain of stars leading from the 5½-magnitude star seems to be attached to this bucket, drawing it up from a celestial well.

And the cluster certainly is beautiful at medium power. Dozens of "bright" suns (the brightest being about magnitude 12.5) burst forth from a seething glow of faint stars that seem ready to boil out of the cluster's core. Indeed, the nuclear region displays a cauldron of bright suns, which at 130×, looks highly fractured. The brightest section is skewed to the northwest, where the stars concentrate in a wedge-shaped pattern. I've

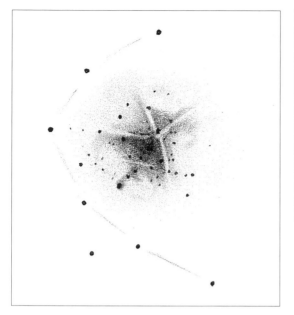

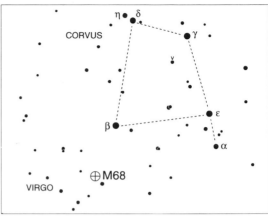

spent hours studying that mysterious center, trying to fathom its complexity. The wait was worth it. Look at the drawing and notice the four darkest lanes (white in this negative rendering), which form a windmill-like formation over the southeastern half. Other dark patches are clearly created by gaps between strings and arcs of bright stars, especially to the south. The "windmill" is but one of many patterns of dark lanes in this region. What you see really depends on how you look and what's on your mind. While your imagination is still engaged, look for a striking detached portion in the northern halo, where you will find a dark "footprint."

M69

NGC 6637
Type: Globular Cluster
Con: Sagittarius
RA: 18^{h}31^m.4
Dec: −32°21′
Mag: 7.7; 7.4 (O'Meara)
Dia: 10′
Dist: 33,600 l.y.
Disc: Nicolas-Louis de
Lacaille, 1752

MESSIER: [Observed 31 August 1780] Nebula without a star in Sagittarius, below the left arm and close to the bow. Nearby there is a ninth-magnitude star. The luminosity is very faint and can be seen only under good conditions, and the slightest illumination of the micrometer's crosshairs causes it to disappear. Its position was determined from (Sagittarii. This nebula was observed by M. de la Caille, and given in his catalogue. It resembles the nucleus of a small comet.

NGC: Globular, bright, large, round, well resolved, stars of 14th to 16th magnitude.

Messier's 69th catalogue entry is one of several tiny objects in "globular cluster alley" – a 10°-long strip of sky along the bottom of the Sagittarius teapot between Epsilon (ε) and Zeta (ζ) Sagittarii that contains three Messier globulars and one NGC globular. M69, a 7th-magnitude globular with a diameter of 10′, is the one closest to Epsilon about 3½° to the north-east. At 23×, it is a uniformly bright fuzzball next to which shimmers an 8th-magnitude topaz star. When I relaxed my gaze, several patches of light in the area pulled my eye away from M69. These patches turned out to be sections of Milky Way sliced into little shards by dark nebulous veins. The most prominent section lies to the southeast, where it forms a bank of patchy light through which a black river flows. All this leads to the diminutive 9th-magnitude globular NGC 6652. NGC 6652 lies about 8′ south of a 6.9-magnitude aqua-tinged star. After looking at these colorful stars, I

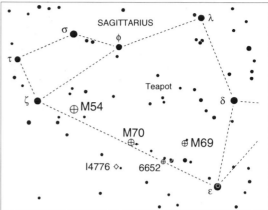

suddenly realized how sickly M69's pallor is. In fact, all the clusters in globular alley remind me of ancient artifacts that have faded and weathered with time, parts of them disintegrating into loose particles. This area is charged with a fair amount a intergalactic dust, which dims the light emanating from these distant objects.

Although moderate power will resolve a few stellar members, the cluster is much more satisfying at 130×. Immediately the core appears asymmetrical to the north – an effect caused largely by a dark rift slicing through the southern half. With a longer gaze the entire cluster fragments into patches. Here we have a patchy cluster mimicking its Milky Way surroundings! In fact, M69 contains too many faint patches of darkness for me to render, though not much detail reveals itself to the east. Can you see the dark lagoon north of the core?

M70

NGC 6681
Type: Globular Cluster
Con: Sagittarius
RA: 18^h 43^m.2
Dec: $-$32° 18′
Mag: 7.8
Dia: 8′
Dist: 35,200 l.y.
Disc: Messier, 1780

MESSIER: [Observed 31 August 1780] Nebula without a star close to the previous one [M69], and on the same parallel. Nearby there is a ninth-magnitude star and four faint, telescopic stars almost in a straight line, very close to one another, and which lie above the nebula, as seen in an inverting telescope. The nebula's position was determined from the same star, ε Sagittarii.

NGC: Globular, bright, pretty large, round, gradually brightening toward the middle, stars from 14th to 17th magnitude.

Although M70 lies at about the same distance as M69 and is roughly the same size (82,000 light years across), it is moving away from us twice as fast. Use binoculars to find the globular's tight, hazy glow about 2° east of M69. Telescopically at low power, the cluster looks like the sizzling end of a fuse of stars to the north. Jones offered a similar metaphor: "As if attached to the cluster, there is a little slightly curved 'tail' of small stars, shooting off like sparks to the NNE. These may be the stars mentioned by Messier." Flammarion spoke of the cluster being decorated with a pretty double star to the northeast, and this star pair is a welcome sight even at high power.

If I didn't know better, I probably would have mistaken M70 for a spiral galaxy. Three distinct "arms" curl out of a very tight, fuzzy core. Most noticeable are the northern and southern arms, both of which are riddled with starlight. The southern string ends abruptly at a single bright star. Can you resolve the cluster's outer halo? I cannot; it remains an elusive

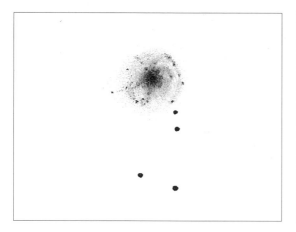

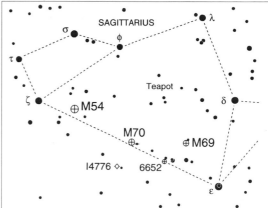

haze with teasingly knotty sections. Use high power and you might have some success. I noted that the core remained unresolved in the 4-inch. It is much more compressed than M69's core and, as Skiff observed, it is slightly elongated. Try comparing M69 and M70 on the same night.

Once when I was looking at M70 with low power, I picked up with averted vision what appeared to be a very faint nucleus to a comet about 1° to the southeast. I checked the object with high power, and it remained an unresolved fuzzy star. When the object did not move after an hour, I suspected it was a planetary nebula, which it turned out to be. It is 10.4-magnitude IC 4776.

M71

NGC 6838
Type: Globular Cluster
Con: Sagitta
RA: 19h 53m.8
Dec: + 18° 47'
Mag: 8.4; 8.0 (O'Meara)
Dia: 7'.2
Dist: 13,000 l.y.
Disc: Probably Philippe Loys de Chéseaux, 1746; Johann Gottfried Kohler observed it around 1775

MESSIER: [Observed 4 October 1780] Nebula discovered by M. Méchain on 28 June 1780, between the stars γ and δ Sagittae. The following 4 October, M. Messier searched for it. The light is very faint and it contains no stars. The slightest illumination causes it to disappear. It lies about 4 degrees below the nebula that M. Messier discovered in Vulpecula; see number 27. It is plotted on the chart for the comet of 1779.

NGC: Cluster, very large, very rich, pretty much compressed, stars from 11th to 16th magnitude.

M71 was once an object of controversy among professional astronomers. Some argued it is a loose globular, while others claimed it's an extremely dense open cluster. There is no doubt today, however, that M71 is a globular – a very near one, being only 13,000 light years distant, which is why it is so easily resolved and appears to lack the dense center typical of more distant globulars.

It is easily spotted midway between the bright stars Gamma (γ) and Delta (δ) Sagittae, and a mere 20' northeast of 9 Sagitta, all in the shaft of the famous Celestial Arrow. Anyone viewing this cluster through a telescope will immediately understand that former classification controversy. Here is a sizable (7') yet compact glow of moderately bright to faint stars (typical of globulars), whose loose center resolves well with medium magnification (typical of open clusters). In a low-power field of view, M71 occupies the center of an oval area outlined by four distinctive Y-shaped asterisms, all oriented in different directions. M71 itself is rather Y-shaped, especially if you include the two 12th-magnitude stars just outside its round and moderately condensed haze. Webb called it an "interesting specimen of the process of stellar evolution" – an interpretation seemingly shared by astronomer Isaac Roberts, who, after imaging the cluster in 1890, commented that the curves in its crowded star regions were suggestive of having been produced by the effects of spiral movements.

Through the Genesis, no central condensation is apparent, even with high power. The cluster does sport an arrow-shaped concentration to the southwest. Dark lanes cut through that region in a gridlike fashion.

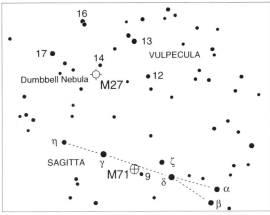

Three of the darkest lanes run southeast to northwest, alternating with waves of starlight flowing from the densest wedge of light to the southwest. Now, relax your mind. Do you see these suns calving away from that wedge like heavy snow down an angled roof? (A chilly thought for warm summer nights, when M71 is best placed in the evening sky.)

Surprisingly, Mallas claimed he could not resolve *any* stars within the cluster in his 4-inch refractor. Skiff believes Mallas possibly assumed the faint stars he resolved near the cluster were field stars, which is usually the case, but here they're actually cluster members. Otherwise, Mallas's observation remains a mystery, since a century before him, Webb saw the cluster "yield to a cloud of faint stars" with high power in a 3.7-inch – presumably a telescope of inferior quality.

M72

NGC 6981
Type: Globular Cluster
Con: Aquarius
RA: 20^h 53^m.5
Dec: –12° 32'
Mag: 9.2
Dia: 6'
Dist: 56,400 l.y.
Disc: Pierre Méchain, 1780

MESSIER: [Observed 4 October 1780] Nebula seen by M. Méchain on the night of the 29–30 August 1780, above the neck of Capricornus. M. Messier searched for it the following 4 and 5 October. Its light is faint like the previous one [M71]. There is a faint telescopic star nearby. Its position was determined relative to fifth-magnitude ν Aquarii.

NGC: Globular, pretty bright and large, round, much compressed in the middle, well resolved.

About 4° (nearly two finger-widths at arm's length) southeast of 3.8-magnitude Epsilon (ε) Aquarii, or about 10° (a fist-width) due east of the naked-eye double star Alpha (α) Capricorni, you come to the faintest Messier globular, M72. At magnitude 9.2 and just 6' in apparent diameter, this globular is easy to pass over. Look for a 9th-magnitude "double star" separated by 5' – the eastern component is, in fact, a star; the western component is M72. Once found, use moderate magnification to enlarge the cluster's disk.

M72 spreads across about 98 light years of space and, like M71, is classified as a very open globular. The cluster's brightest members are near the limit of resolution in small apertures. Some 14th-magnitude suns can be detected in the outer envelope at 72×, but only with difficulty: look for a pair of stars to the south, and another, closer pair to the northeast. D'Arrest partly resolved the cluster at 9× and found it well resolved at 123×; though his observation might have been made with an 11-inch refractor.

Although M72 is still difficult at 130× in the 4-inch, I can resolve a fair number of members and trace some faint "arms." But this is by no

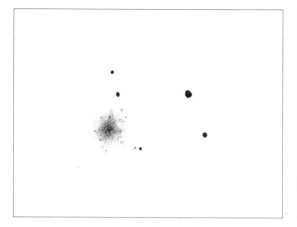

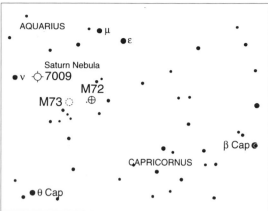

means a simple task! By resolving the stars, I mean being able to see insignificantly faint speckles pop in and out of view in an otherwise foggy moor of starlight. Photometry of this globular reveals that its brightest star shines at magnitude 14.2. The core, which is apparently loose when viewed through large instruments, looks slightly diamond shaped and impenetrable through the 4-inch.

It's not that M72 is a *bad* cluster, it is just better suited for large telescopes and high magnifications. Perhaps many millennia in the future, as the cluster races toward us at 160 miles per second, small-telescope users won't find the object as difficult to observe.

M73

NGC 6994
Type: Asterism?
Con: Aquarius
RA: 20^h 58^m.9
Dec: –12° 38′
Mag: 8.9
Dia: 1′.4
Dist: Unknown
Disc: Messier, 1780

MESSIER: [Observed 4 and 5 October 1780] Cluster of three or four faint stars, which, at first glance, resembles a nebula, and which does contain some nebulosity. This cluster lies at the same declination as the previous one [M72]. Its position was determined using the same star, ν Aquarii.

NGC: Cluster, extremely poor, very little compressed, no nebulosity.

"A trio of 10 mag. stars in a poor field" is how an unimpressed Smyth described this unremarkable M object in his *Cycle of Celestial Objects*. A Y-shaped asterism conveniently situated 1½° east of M72, M73 is actually a grouping of four stars with no apparent connection, except that they lie in the same line of sight. Some sources still list it as an open cluster, but probably it is not. Such a chance grouping of four 10th-magnitude stars is not that improbable. Indeed, in his *Atlas of Deep-Sky Splendors* Hans Vehrenberg notes that asterisms of this nature can be found in any part of the sky. However, no detailed measurements have apparently been made of these stars, so the jury is still out on whether they are in some way related.

I enjoy the shape and texture created by these four stars, because, if nothing else, they incite mental flights of fancy. Once, I brought a tape of Gustav Holst's *The Planets* with me into the field and played *Venus* as I stared into the eyepiece. The Y-shaped asterism, which is aligned east–west, suddenly transformed into a Flash Gordon-style rocket ship sailing through interstellar space. Because the stars gradually fade to the west, the ship is seen obliquely from behind. The two bright end stars of the Y are the burning rockets. It's a stretch, granted, but for this object in particular a little flash is just what's needed!

I do not experience the same thrill of discovery that Messier must have felt when he encountered these stars, because even at 23× in the 4-inch M73 does not look fuzzy – in the sense of being cometlike, or nebular, as Messier described it. Obviously, the optics of Messier's telescope were not of the quality of optics in modern telescopes, and that is probably one

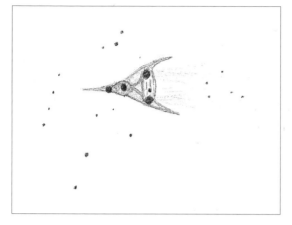

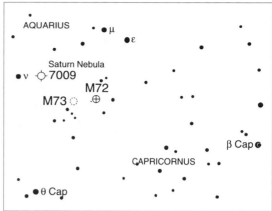

reason why this grouping looked fuzzy to him; another reason could have been poor atmospheric conditions. So I went in search of another, more cometlike asterism suitable for the 4-inch. And I found one! Try sweeping your telescope at low power about 1° to the southwest of M73. Do you encounter a small haze? It is yet another asterism of four stars, in the shape of a Y, only this one is a mirror-image of M73. These stars are closer together than those in M73, giving them a nebular appearance at 23×. Archinal independently chanced upon these stars as well, and noted their similarity to the M73 grouping. This also corroborates Vehrenberg's comment that asterisms of this nature are not uncommon!

For another challenge, sweep your telescope 2° northeast of M73 and look for a slightly swollen 8th-magnitude "star." This is the Saturn Nebula, a pale green planetary nebula. A 12.8-magnitude star blazes at the center of the nebula's 44″×23″ disk. Use high magnification to see that central star.

I once had an opportunity to view this nebula through the great 60-inch reflector atop Mount Wilson in California. On that fine, moonless night in August 1986, the planetary displayed two sharply defined green oval disks surrounding the central star, which burned brightly at 400×. On closer inspection, each ring had a series of condensations highlighting it and two long rays with tight knots stretched beyond the ansae of the rings. The nebula bore a close resemblance to the planet Saturn seen edgewise, with bright moons adorning the tips of the hairline rings.

M74

M74

The Phantom
NGC 628
Type: Spiral Galaxy
Con: Pisces
RA: 1ʰ 36ᵐ.7
Dec: +15° 47′
Mag: 9.4; 8.5 (O'Meara)
SB: 14.4
Dim: 10′.5 × 9′.5
Dist: 32 million l.y.
Disc: Pierre Méchain, 1780

MESSIER: [Observed 18 October 1780] Nebula without a star, close to star η in the ribbon of Pisces. Seen by M. Méchain at the end of September 1780, and about which he reported "This nebula does not contain any stars. It is quite broad, very dim, and extremely difficult to observe; it may be distinguished more accurately during fine frosts." M. Messier searched for it and found it to be as M. Méchain described. It was directly compared with the star η Piscium.

NGC: Globular cluster, faint, very large, round, pretty suddenly much brighter toward the middle, some stars seen.

Blazing with the light of 40 billion suns, flinging spiral arms across 97,000 light years of space, M74 is a prima donna among open-face spiral galaxies. At least, that is how it comes across in long-exposure photographs taken through large telescopes. In small telescopes, it is more like a phantom, which is the nickname I have given it. No object in the Messier catalogue has proven more troublesome, more elusive, more provocative to amateur astronomers than this giant spiral. The problem is that the galaxy's large apparent size (10′.5) and very low surface brightness require a very dark sky for it to be seen well, if at all! Méchain was right on the mark when he said, "This nebula . . . is quite broad, very dim, and extremely difficult to observe; it may be distinguished more accurately during fine frosts." The fine frosts he refers to, I am sure, are those incredibly transparent evenings following the passage of a cold front, when the night sky is free from moisture and atmospheric contaminants, and the stars can be seen with crystal clarity away from city lights. On these nights you have the best chance to see dim and diffuse objects.

I searched for M74 many nights with Harvard Observatory's 9-inch refractor without success. The fact is, so large an aperture under less-than-perfect skies doomed my quest before it even started. It's best to use a small-aperture instrument, low power, and a wide field of view on the finest of nights.

Locating the field is easy. Look about 1½° east-northeast of 3.6-magnitude Eta (η) Piscium. From dark skies 23 × shows M74 as an obvious but pale disk of uniform light. The longer you look, the more detail you should see. Watch how the core slowly materializes into a compact orb punctu-

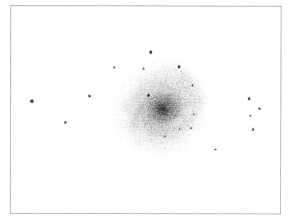

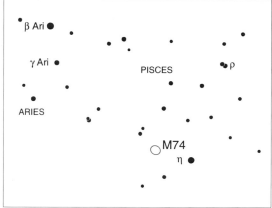

ated by a pinpoint of light. With averted vision a diffuse outer skirt also takes shape. Jones believed that M74 could not stand anything but low power, but I disagree. If you can make out the galaxy's inner core, then that is the time to change to higher power. The core itself is mottled with the light of many dim stars, which emerge only with high powers. It is this chance alignment of field stars projected against the nucleus that obviously caused John Herschel to mistake M74 for a globular cluster, which in turn led Dreyer to list it as a globular in the *New General Catalogue* (see the *NGC* description earlier).

M74 is also a very difficult galaxy to sketch. Its spiral patterns are like spirits that weave in and out of view. So fleeting are these moments that I needed hours (and nights) of observing to confirm and reconfirm what I thought I was seeing. It is ironic that once I identified the field around M74 in the telescope, I used 7 × 35 binoculars to spot the galaxy. Imagine, here is the object I couldn't see with a 9-inch refractor from Cambridge, Massachusetts, yet here in my sparsely populated corner of Hawaii, it is visible in 7 × 35s. This is an excellent example of the degrading effect of light pollution.

M75

NGC 6864
Type: Globular Cluster
Con: Sagittarius
RA: 20^h 06^m.1
Dec: –21° 55′
Mag: 8.6
Dia: 7′
Dist: 59,300 l.y.
Disc: Pierre Méchain, 1780

MESSIER: [Observed 18 October 1780] Nebula without a star between Sagittarius and the head of Capricornus. Seen by M. Méchain on 27 and 28 August 1780. M. Messier searched for it the following 5 October, and on the 18th compared it with sixth-magnitude Flamsteed 4 Capricorni. It seemed to M. Messier that it consists only of very faint stars, but contains some nebulosity. M. Méchain described it as a nebula without stars. M. Messier saw it on 5 October, but the moon was on the horizon, and it was not until the 18th of the same month that he was able to make out its form and determine its position.

NGC: Globular, bright, pretty large, round, very much brighter toward the middle to a much brighter nucleus, partially resolved.

Messier's 75th deep-sky curiosity is also one of the more challenging to find, because it lies in a southern region of sky devoid of bright guidepost stars. This isolated, southern cluster, one of the more distant of the Messier globulars at more than 59,000 light years, resides 8° southwest of 3rd-magnitude Beta (β) Capricorni. To reach it, if your telescope has setting circles, you could start at Pi (π) Sagittarii in the spoon asterism northeast of the teapot, then move the telescope 1° south in declination and about 1 hour (14°) east in right ascension; but it is more fun just to sweep. You will find the cluster nestled in a faint asterism that bears a striking resemblance to the constellation Scorpius – but it is a backwards scorpion. Once thought to be a fugitive globular fleeing the bounds of our galaxy, M75 is, on the contrary, a healthy member of the Milky Way's halo community; in fact, its distance pales in comparison with that of the truly faraway globulars, such as NGC 2419 in Lynx, which is an incredible 380,000 light years distant! Despite its great distance, M75's highly compact 7′ disk shines at a respectable magnitude 8.6.

At 23 × I swept over the cluster a few times before sighting its slightly swollen starlike disk. At moderate power the cluster looks larger but remains largely unresolved, though a few outlying 14th-magnitude field stars do start to materialize. Messier claimed to have resolved this cluster,

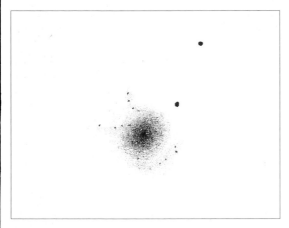

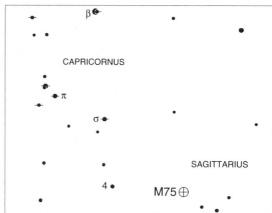

but one has to wonder, given that he failed to resolve stars in the brighter giants, like M13 and M5!

At 72× the globular shows three tiers of brightness: a starlike nucleus burns at the heart of a dense vaporous shell, which is surrounded by a round outer envelope. High power brings out at least three "arms." The most prominent one to the northeast looks like a band of faint starlight drifting slowly away from the cluster, like renegades fleeing from the fugitive pack.

M76

Little Dumbbell Nebula
NGC 650 and NGC 651
Type: Planetary Nebula
Con: Perseus
RA: $1^h 42^m.4$
Dec: $+51° 34'$
Mag: 10.1
Dia: 67″
Dist: 3,900 l.y.
Disc: Pierre Méchain, 1780

MESSIER: [Observed 21 October 1780] Nebula near the right foot of Andromeda, seen by M. Méchain on 5 September 1780, and which he described thusly: "This nebula does not contain any stars; it is small and faint." The following 21 October, M. Messier searched for it with his achromatic telescope and it seemed to him to consist of very faint stars with some nebulosity, and the slightest illumination of the micrometer crosshairs caused them to disappear. The position was determined from fourth-magnitude φ Andromedae.

NGC: NGC 650 and 651, both very bright, are the preceding and following components of a double nebula.

Smaller, fainter, and less popular than its cousin the Dumbbell Nebula (M27) in Vulpecula, the Little Dumbbell in Perseus is nonetheless a very dramatic planetary. Sometimes, unfortunately, the popularity of an M object seems to be based more on how bright or large it appears in the night sky than on how much detail it reveals through the telescope. But in the case of M76, its beauty lies not in its visual punch, but in its wealth of subtle detail, which lures small-telescope users into a web of visual sug-

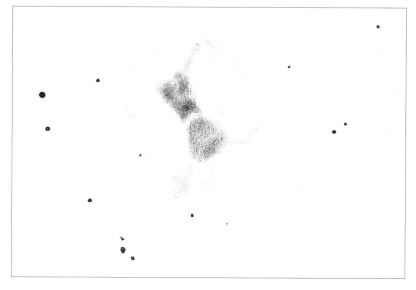

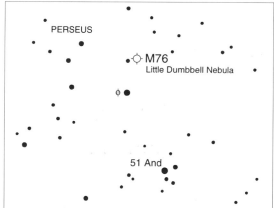

PERSEUS

◇ M76
Little Dumbbell Nebula

φ ●

51 And

gestions. I'd place it among the most surprising and mysterious objects in the Messier catalogue for viewing with backyard telescopes.

Certainly astronomers at the turn of the century must have been awed by its image in long-exposure photographs. Instead of a ghostly ring or disk of light encircling a central star – as is characteristic of planetaries – here was a 16th-magnitude central star in a *rectangular bar* of light made up of two prominent nebulous patches (thus the double *NGC* entry) flanked by an irregular array of luminous knots and nebulous arcs. In 1891 Isaac Roberts suggested that M76's unusual character is due to our seeing its broad ring of material edgewise. But the object's complex appearance might derive more from the way gases of varying densities near the central star illuminate and obscure the star's light.

M76 is about five times more distant than M27, but they are about the same actual size, so M76 only *appears* smaller. Some catalogues severely underestimate the brightness of the Little Dumbbell, listing it as 12th magnitude, which is much too faint. Indeed, as proof, M76 is at the limit of detectability in 7 × 35 binoculars, which is about 10th magnitude. At 23× in the 4-inch, its small, opalescent disk reposes in a rich field of faint stars, many of which are pairs, less than 1° north-northwest of yellowish Phi (φ) Persei. To avoid passing over it with low power, sweep slowly and use averted vision. With peripheral vision the planetary seems to swell. This is not an illusion. Your eyes are picking up the faint light from the nebulous loops that connect to the four corners of the brighter rectangular bar.

At 72× I found the Little Dumbbell to resemble the Crab Nebula (M1), probably because both objects display bright patches of nebulosity separated by a dark lane and bordered by misty frills. And these details are immediately apparent in M76. Despite its irregular appearance in photographs, the planetary looks symmetrical through small-aperture telescopes; there is an hourglass-shaped inner nebula, oriented northwest to southeast, with two semicircular arcs, one to the southeast, the other to the northwest – much like M27!

Interestingly, when I look at M76 with moderate power, the southwestern loop seems more pronounced, but at high power the northeastern one does. Although I can see the arcing extensions attached to each corner of the rectangle, the brightest sections (knots) within each loop are clearly separated from the hourglass. The hourglass itself is made up of bright knots, which could be faint stars superposed on the nebula. The southwestern half is the brighter. Rosse also detected "subordinate nodules and streamers," which led him incorrectly, though understandably, to deduce that this was a spiral nebula. There appears to be a hint of brightening *near* the center of the rectangle, but do not confuse this isolated glow for the 16th-magnitude central star. As the drawing shows, M76 has a heart of darkness (light in the negative rendering), and the central star hides in the deep recesses of that bleak cavity.

M77

NGC 1068
Type: Spiral Galaxy
Con: Cetus
RA: $2^h 42^m.7$
Dec: $-0° 01'$
Mag: 8.9
SB: 13.2
Dim: $7'.1 \times 6'.0$
Dist: 47 million l.y.
Disc: Pierre Méchain, 1780

MESSIER: [Observed 17 December 1780] Cluster of faint stars, which contains nebulosity, in Cetus, and at the same parallel as the star δ, which is reported to be of third magnitude, but which M. Messier estimates to be only of fifth magnitude. M. Méchain saw this cluster on 29 October 1780 as a nebula.

NGC: Very bright, pretty large, irregularly round, suddenly brighter toward the middle, some stars seen near the nucleus.

This object is the prototype of a peculiar class of extragalactic objects known as *Seyfert galaxies*. These systems have very active nuclei, which are potent emitters of radio-wavelength energy and whose spectra show strong emission lines – characteristics also displayed by the distant quasistellar objects known as *quasars*. But quasars are infant systems billions of light years away, whereas energetic M77 is a mere 47 million light years distant. Apparently, gas clouds (some with the mass of 10 million suns) are blasting away from the nucleus of M77 with velocities up to 360 miles per second and enough energy to power several million supernovae explosions! M77 is the only Seyfert galaxy in the Messier catalogue and is the closest you will come to seeing a quasar in action.

And, indeed, the nucleus of M77 is bright. You can see it shining like a 10th-magnitude star about 1° southeast of 4th-magnitude Delta (δ) Ceti, just west of a real 10th-magnitude star; so at low power, and at first glance, the two look like a fine double star. But soon you will notice the faint halo of light surrounding M77, whose contribution raises the galaxy's brightness to 9th magnitude. You will need moderate or high magnification to see the galaxy's peculiar arms, which long-exposure photographs reveal as tight and knotty close to the brilliant nucleus but fainter and less tightly wound farther out. Burnham notes that the knots of the inner arms can be glimpsed through a 4-inch telescope on the finest nights, and he is absolutely right! In fact, these knots are what caused earlier astronomers like the Herschels and Smyth to perceive M77 as a resolved star cluster. Rosse noted it as a blue spiral. Can you detect any color?

After observing the galaxy on three different nights with the 4-inch at high power, I traced its inner spiral patterns with a high degree of confidence and accuracy. A tiny but strong bar slices across the nucleus in an east–west direction and quickly curves away into tight but graceful arms. In 1993 infrared images of M77 made at Kitt Peak National Observatory revealed a nebulous bar extending through the nucleus. The question is whether this infrared bar matches the bar seen visually. Most perplexing, at first, is the galaxy's northern arm, which appears to branch from a knot close to the nucleus. You really have to study this area long and hard before you can begin to distinguish the individual arms from the nucleus. The difficulty is that these arms, like those of M66, are broken into knots and patches, so you have to find a way to piece the chaos together into a coherent picture. The trick is to use as much power as possible, to hyperventilate to heighten visual acuity (see page 00), and to play with the focus and use averted vision until you can cause the features to blend. Can you resolve the faint knot to the east of the nucleus, in the innermost spiral?

Returning to Delta Ceti, look about 6° southwest, where you'll find the fascinating variable star Omicron (o) Ceti, also known as Mira. David Fabricius (1564–1617) is credited with the discovery of Mira in August 1596, though Korean and Chinese astronomers may have noticed this "guest star" several months before him. This winking star fades in and out of naked-eye visibility like clockwork every 333 days. It fluctuates in brightness from a maximum of magnitude 3.4 (usually) to a minimum of 9.3. However, it has peaked as bright as magnitude 0.9. William Herschel made that historic observation in November 1779, when he found that Mira "excelled Alpha Arietis so far as almost to rival Aldebaran; and continued in that state a full month." The star varies in brightness because it has entered the pulsating, red-giant stage of its life cycle. Mira has roughly the sun's mass, so by studying its behavior, we can learn more about how our own sun will act once it enters old age.

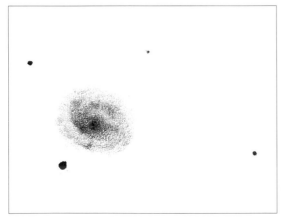

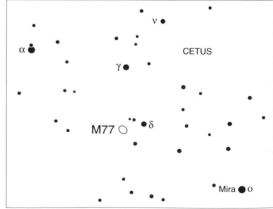

M78

NGC 2068
Type: Diffuse Nebula
Con: Orion
RA: 5^h 46^m.7
Dec: + 0° 03′
Mag: 8.0 (O'Meara)
Dim: 8′ × 6′
Dist: 1,630 l.y.
Disc: Pierre Méchain, 1780

MESSIER: [Observed 17 December 1780] Cluster of stars, with a lot of nebulosity in Orion, and on the same parallel as δ in the belt, which was used to determine its position. The cluster crosses the meridian 3° 41′ after the star and is 27′ 7″ higher in the sky. M. Méchain saw this cluster at the beginning of 1780, when he reported it as follows: "On the left-hand side of Orion, 2 to 3 minutes in diameter. Two fairly bright nuclei are visible, surrounded by nebulosity."

NGC: A large, bright wisp, gradually much brighter toward a nucleus, three stars involved, mottled.

Before beginning this book, I had looked at M78 only once. Some 20 years ago Peter Collins, a visual discoverer of four novae, had turned the Harvard College Observatory's 9-inch Clark refractor to the nebula, and I took a peek. Once home, I looked it up in *Burnham's Celestial Handbook*, which supported my view of very little detail. Now, having spent many evenings with it using a much smaller telescope, I realize my misfortune

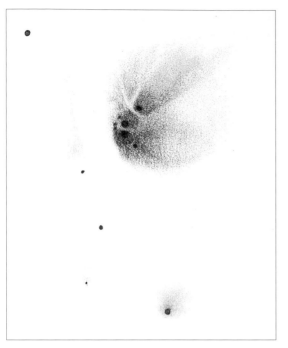

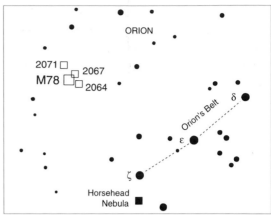

in neglecting this wonderfully mysterious object (more on the mysteries later).

Located about 2½° northeast of Zeta (ζ) Orionis, M78 is far enough away from the nebular madness surrounding Orion's Belt and Sword to be easily overlooked. In fact, M78 is the brightest of three specterlike glows in the region. Its companion nebulae, NGC 2067 and NGC 2071, are described subsequently. All three shine by reflected light from hot, young Type-*B* stars and are part of a greater complex of nebulosity sweeping through much of Orion. Most sources don't publish a magnitude for M78, even though it is visible with 7 × 35 binoculars. The reason is that there are very few good measurements of nebula magnitudes. My magnitude estimate of 8.0 was made with binoculars. If you have good to excellent eyesight, challenge yourself by trying to detect it without aid.

A close pair of roughly 10th-magnitude stars burn at the heart of M78. Even at low power, they look like bloodless eyes peering back at you through a frosty window. Smyth and Webb described M78 as a wispy nebula, and Rosse saw spiral structure. Today its appearance is generally compared to a comet's. Indeed, the nebula does resemble a comet with a split nucleus, parabolic hood, and a tail that fans from south to east. (Anyone who has seen the sunward fan of Periodic Comet Encke will appreciate how closely M78 resembles that feature.)

But there are "comets" within this "comet." Most striking is the long tail of material flowing southeast of the southern nucleus. Look at this tail under magnification and you will see it has two parts: a short, stubby fan associated with one of the 10th-magnitude stars and a long, graceful tail associated with a 13.5-magnitude. The nucleus's northern component has a roughly 14th-magnitude companion to the east, around which diffuse material streams out to form a "flame" of nebulosity.

With high magnification focus your gaze immediately west of the two 10th-magnitude nuclei. At first you will see what appears to be a star, then two stars. But are these really stars? Perhaps they are knots of nebulosity. There is yet another very faint star or knot immediately to the southwest of the southern component. No image I have seen shows the inner details of M78, so it has been difficult to confirm these sightings. Perhaps someone with a CCD camera and an image-processing program with unsharp masking can bring out these details.

As I alluded to earlier, M78 is not without certain mysteries. When I first looked at it through the eyepiece, I expected, based on several written accounts, to see nebulosity surrounding two bright stars. But there were *three* obvious stars (as noted in the *NGC* description): the two dominant 10th-magnitude stars and a fainter one. Then I read in Luginbuhl and Skiff's book that the third star shines at magnitude 13.5. That seemed to match my visual impression, but when I returned to M78 a month later, the third star appeared fainter than 13.5, perhaps by as much as a magnitude. In a 1975 *Astrophysical Journal* paper, the star was measured to be magnitude 13.1. Could it be a variable star?

Here's something else to investigate. The nebula has a very faint, southern extension, superposed on which appears to be a vein of dark nebulosity forming what I call the "skeleton's wrist." Actually, the wrist itself is not difficult to make out; it is the lane of darkness that separates the two southeast "comet tails." The wrist, however, then branches into two extremely tenuous dark wisps: one cuts sharply to the south, the other, more difficult extension, goes southwest – the skeleton's thumb and forefinger. I saw these other features only once in several sessions. Can you see them?

While you are in the neighborhood of M78, check out these interesting NGC objects. Refer to the finder chart for their locations. NGC 2071 to the northeast is an easy nebulosity, though it has a lower surface brightness than M78. At first glance the nebula appears round, but it is actually extended to the south and slightly east. See if you can detect it in binoculars. More intriguing is NGC 2067, which I noticed only by circumstance. On one very gusty night, the Genesis was jittering slightly, setting the field in motion enough for my peripheral vision to catch sight of sharp-edged NGC 2067 to the northwest. There is yet another nebulosity – the minuscule (1′.5×1′) NGC 2064 to the southwest, which I have only suspected. Thus it is not included in the drawing.

M79

NGC 1904
Type: Globular Cluster
Con: Lepus
RA: 5ʰ 24ᵐ.2
Dec: –24° 31′.5
Mag: 7.7
Dia: 6′
Dist: 43,000 l.y.
Disc: Pierre Méchain, 1780

MESSIER: [Observed 17 December 1780] Nebula without a star lying below Lepus, and on the same parallel as a sixth-magnitude star. Seen by M. Méchain on 26 October 1780. M. Messier looked for it the following 17 December. This is a fine nebula; the center is bright, the nebulosity slightly diffuse. Its position was determined from fourth-magnitude ε Leporis.

NGC: Globular cluster, pretty large, extremely rich and compressed, well resolved into stars.

The next time you are under a very dark sky in a southerly locale, try the following. Using only your eyes, visualize a line from Alpha (α) to Beta (β) Leporis, and extend it southwest about 4½° (about the width of two fingers held at arm's length). There you will see a 5th-magnitude star. If you are properly dark-adapted, you might see, with time, what appear to be two very faint stars nearby. One is about ½° to the southwest and indeed is a 7.5-magnitude star. However, the other, about ½° to the northeast, is glob-

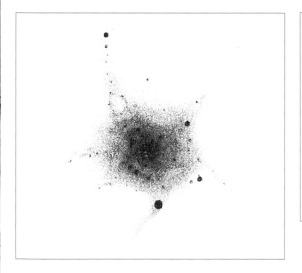

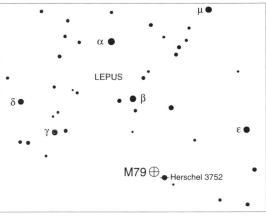

ular cluster M79. (To do this properly, you will need to spend 15 minutes or more trying. But do try, because I certainly surprised myself by seeing it!)

Needless to say, M79 is an easy target for binoculars – a 7.7-magnitude fuzzball bracketed nicely by two 9th-magnitude stars. Even in the 4-inch at low power, it is surprisingly small and tight, so tight that I once swept over it, mistaking the globular for a star. It is hard to believe that this seemingly insignificant patch of light spans some 75 light years of space. Then again, if you consider how rapidly the globular is receding from us – at about 143 miles per second – it is easy to imagine it diminishing from view, thus justifying its size in our sky.

At moderate power M79 has a distinctive starfish shape that is punctuated on its northern fringe by a 12th-magnitude star. I recall, as a teenager, seeing this shape through my 4½-inch reflector and lightheartedly thinking how much it resembled a fly in a melting ice cube. One frigid evening many years later, as Lepus hopped above some distant fir trees at Oak Ridge Station in Harvard, Massachusetts, I turned a 16-inch reflector onto the globular and resolved it right to the core.

Even in a 4-inch telescope, M79 is a very satisfying cluster. It contains a smattering of fairly bright members against a blazing core, which was first described by both Smyth and Webb in the nineteenth century. The cluster's boxy sides are obvious in both the inner and outer regions. The outermost reaches can be resolved at moderate power, even though the individual stars shine between 13th and 14th magnitude. The tightly wound center is a demon for small apertures. Any proper attempt at probing its depths requires high magnification, keen averted vision, and lots of patience.

If you spend enough time studying the nuclear region with high power, you might notice an asymmetry. That is, the bright western side fades gradually and subtly toward the east. Part of the asymmetry results from the presence of a nice string of glittering stars on the northwest side of the nucleus. Furthermore, there's a gap between them, resulting in fewer stars on that side of the cluster. Two smaller stumps of starlight to the west and southwest add further lopsidedness to the interior.

The cluster also shows five "arms" extending in various directions. At first they will look fuzzy, but with increased magnification, they can be resolved into faint streams of stars. Most intriguing is the southwestern arm, which seems to consist of a series of thinning stellar arcs that drip toward a bright star (remember the melting ice cube?).

As the photograph shows, all these arms develop into crazy patterns of extremely faint outlying stars. I have yet to see these wild swirling patterns in the 4-inch. Perhaps one way to see them would be to use very high magnification and focus on each arm individually. I expect each search would consume hours. But if you feel up to the task, you could piece together a drawing that would rival the best photographs.

M80

NGC 6093
Type: Globular Cluster
Con: Scorpius
RA: 16^h 17^m.0
Dec: –22° 59′
Mag: 7.3
Dia: 9′
Dist: 27,000 l.y.
Disc: Messier, 1781

MESSIER: [Observed 4 January 1781] Nebula without a star in Scorpius, between the stars *g* [now ρ Ophiuchi] and δ. It was compared with *g* to determine its position. This nebula is circular; the center is bright and resembles the nucleus of a small comet, surrounded by nebulosity. M. Méchain saw it on 27 January 1781.

NGC: Remarkable globular cluster, very bright, large, very much brighter in the middle, readily resolved, contains stars of the 14th magnitude and fainter.

In an article titled "An Opening in the Heavens" in the 1785 edition of *Philosophical Transactions,* William Herschel shared his impressions of a 4°-wide swath of dark nebulosity above the Scorpion and its fanciful relationship with M80: "It is remarkable that [M80], which is one of the richest and most compressed clusters of small stars I remember to have seen, is on the western border of it and would almost authorise a suspicion that the stars of which it is composed, were collected from that place and had left the vacancy." That vacancy notwithstanding, M80 in fact lies in a very busy region on the western bank of the Milky Way band. It is conveniently placed about 4½° northwest of the ruddy 1st-magnitude star Antares, or Alpha (α) Scorpii – about halfway between Antares and Beta (β) Scorpii. Like M79, M80 is a tiny globular cluster with an extremely dense core and

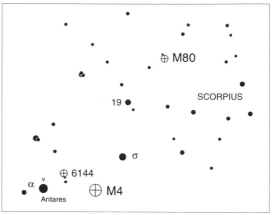

a spherical halo. The owner of a rich-field telescope will be truly rewarded by the sight of M80's thick pack of 7th-magnitude starlight glistening from amidst the galactic pandemonium.

At 23× in the 4-inch the core seems to glow with a faint yellowish light that gradually fades to a silver sheen in the outer envelope. Because the cluster appears so tiny – just 9′ in apparent diameter – I immediately switch to high power, which starts to resolve some cluster members, even though the brightest stars shine between 13th and 14th magnitude. The outer hood seems to scintillate, like flecks of mica. The high density of the cluster makes it difficult to resolve individual stars; high power is a must. At 130×, the core appears fractured, broken into at least three bright shards separated by delicate dark rifts.

In 1860 a 7th-magnitude nova (known as T Scorpii) flared from the center of M80. The new star outshone the globular itself for a few days before it rapidly faded into obscurity. T Scorpii was one of only two (or perhaps three) novae that have been observed in a globular cluster. I like to imagine that M80's fractured core is the tragic result of that energetic blast – a crystalline sphere shattered by powerful shock waves. Now look closely at the nucleus. Does the brightness seem to favor the northeast, where one broad river of stars merges with another? The challenge is to see the dark bay between the rivers.

M81

NGC 3031
Type: Spiral Galaxy
Con: Ursa Major
RA: $9^h 55^m.6$
Dec: $+69° 04'$
Mag: 6.9
SB: 13.0
Dim: $26'.9 \times 14'.1$
Dist: 4.5 million l.y.
Disc: Johan Elert Bode, 1774

MESSIER: [Observed 9 February 1781] Nebula close to the ear of Ursa Major, on the parallel of star d, which is of fourth to fifth magnitude. Its position was determined from this star. This nebula is slightly oval, the center is light, and it is easily visible with a simple three-and-a-half-foot refractor. It was discovered from Berlin by M. Bode on 31 December 1774, and by M. Méchain during August 1779.

Unquestionably, this is the most popular close pair of galaxies in the heavens, and they hold a special place in my heart. Whenever I see M81 and M82, my mind conjures up pleasing memories of two warm summer nights spent under the stars with close friends. One evening, about a decade ago, nova discoverer Peter Collins and I were strolling down a dark country road in Concord, Massachusetts – an old haunt of literary legends Thoreau, Emerson, and Alcott. As we approached a dark forest of oak and elm, we noticed the Great Bear just lurching over the treeline. A minute of quiet contemplation passed before Peter raised his binoculars, which he almost always had handy, and trained them on M81 and M82. I had always thought them to be only telescopic objects. Yet Peter seemed to be giving them an appreciative stare. He then handed me the glasses. Sure enough, there they were, glowing weakly like two spirits fading into the distance.

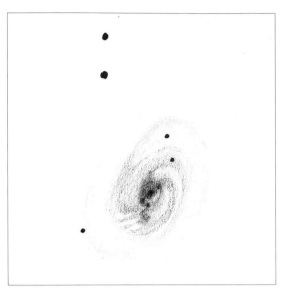

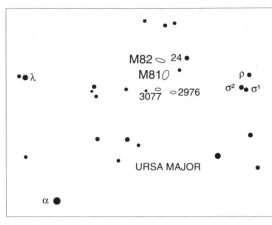

NGC: A remarkable object, extremely bright and extremely large, extended in position angle 156°. It increases in brightness inward, first gradually and then suddenly to a very much brighter center. Bright nucleus.

He then asked if I could see them without the binoculars. Impossible, I thought, so I didn't give it much effort.

That night I did not see the galaxies without aid. But here in Hawaii, after years of trying from other locations, I have finally succeeded with M81. In fact, I estimated the galaxy's brightness with the naked eye to be magnitude 6.8 – the exact figure listed by Luginbuhl and Skiff! Now I have learned that Archinal independently sighted this galaxy without optical aid; he did it while on vacation under dark western skies at high altitude (12,000 feet). I hope these observations inspire you to try to broaden your naked-eye horizon by seeking out this Milky Way neighbor. Some catalogues underestimate the galaxy's brightness by a magnitude or more. And as for its distance, M81 and the equally distant galaxy M83 in Hydra are the farthest objects detectable with the naked eye in nature. At a distance of 4.5 million light years, M81 is twice as far away as both the Triangulum Galaxy (M33) and the Andromeda Galaxy (M31).

M81 and M82 lie about 2° east of 24 Ursae Majoris – a 4.6-magnitude star about a fist-width northwest of Alpha (α) Ursae Majoris. If you live under a dark sky, use binoculars first to locate the galaxy pair. The telescopic view is splendid: M81's egg-shaped, 6.8-magnitude oval, about the apparent size of the full moon, sailing past the smaller, fainter (8.4-magnitude), cigar-shaped ellipse of M82. These two galaxies, separated by a mere 38′, voyaged past one another about 200 million years ago with ominous consequences (the encounter is described a little later); now the galaxies are moving apart. Indeed, although they appear to be cosmic neighbors, M82 is 12.5 million light years more distant than M81.

M82

NGC 3034
Type: Irregular Galaxy
Con: Ursa Major
RA: 9ʰ 55ᵐ.8
Dec: + 69° 41′
Mag: 8.4
SB: 12.8
Dim: 11.2′ × 4′.3
Dist: 17 million l.y.
Disc: Johan Elert Bode, 1774

MESSIER: [Observed 9 February 1781] Nebula without a star, near the previous one [M81]. Both appear in the same telescopic field, the latter is less conspicuous than the former. Its light is faint and elongated. At its eastern end there is a telescopic star. Seen from Berlin by M. Bode on 31 December 1774, and by M. Méchain during August 1779.

NGC: Very bright and very large. An extended ray.

At 23 × in the 4-inch M81 looks enormously larger and more robust than M82. Immediately noticeable is M81's needle-sharp nucleus, which shines with a pale yellow light. Two 11th-magnitude stars burn just south of the core and can easily be mistaken for supernovae. Ironically, in 1993, a supernova in M81 blazed to prominence just west of these stars. Low power also reveals a mysterious "bar" of light crossing the nucleus, but this feature is probably nothing more than an enhancement of the spiral arms seen obliquely along the galaxy's major axis. Indeed, with a prolonged look the bar seems to point toward M82, which can force you to look at that neighbor galaxy. When I do that my averted vision suddenly picks up swooping spiral arms on either side of M81! This is the magic of peripheral vision.

Something odd happens when I change to 72 ×. M82 increases in grandeur, while M81 loses some of its luster (because its faint outer arms have been overmagnified). But the biggest surprise comes at 130 ×, when M81's core is transformed into a misty spring of light caressed by dark vapors, especially to the southeast where a prominent *lash* of darkness abuts this region. Is this a paler version of M64's "black eye"? The bright bar slicing through the nucleus is now contained in a tiny inner ring of nebulosity, the southwestern half of which appears brighter than the northeastern half. Delicate wisps of spiral arms surround the core, and together they look like a still photograph of a rotating lawn sprinkler. The most difficult detail to coax out (but the most rewarding once you see it) is the feathered texture of the spiral arm that lies midway between the outer, northeast arm and the burning core. This texture shows best with moderate power. Thus, to pick out the finest details of M81, you really have to study the outer arms at low power, the core at high power, and the middle arms at moderate magnification.

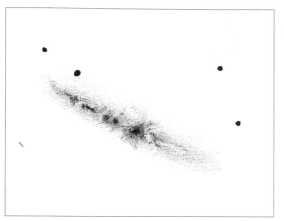

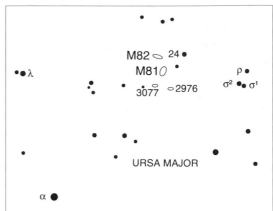

The highly disheveled appearance of M82 stands in marked contrast with its properly groomed spiral neighbor. Here is an extragalactic radical, with spiked "hairs" bristling off a cigar-shaped body tattooed with dark matter. A very unusual galaxy, M82 appears in red-sensitive photographs to have a midsection that is bursting at the seams. Long filaments stream out at right angles from the galaxy's central region; the filaments can be traced out about 34,000 light years; the galaxy itself is only 55,000 light years long! Although M82 was once believed to be experiencing a violent explosion, astronomers now believe its central region is the site of intense starburst activity – containing perhaps 40 or so supernovae in the early stages of expansion. The first starburst episodes probably began 40 million years ago, after M81, which is 10 times more massive than M82, ploughed past M82 like a speeding truck whizzing past a bicyclist. During that encounter, a strong wake of gravity would have smashed into M82, causing interstellar clouds to collapse and trigger new star formation. M82's disrupted appearance probably results from interstellar material being gravitationally dragged away from its disk during that encounter. Newly formed supernovae, which usually flare up during episodes of star formation, could have also blasted material out of the galaxy's plane. As M81 sailed farther away from M82, the jostled interstellar medium of M82 would have gradually fallen back onto its parent galaxy, triggering yet more starburst episodes, which we may still be witnessing today.

At moderate power in the 4-inch M82 is a stunning sight. It reminds me of a ghostly starship, cracked and floating through an interstellar graveyard. As Luginbuhl and Skiff note, the western half of the galaxy is distinctly brighter than the eastern half, though the brightest segment of the fragmented hull lies just east of a central darkening. The far eastern half fades ever so gradually, and it looks as if the galaxy is slowly vanishing before your eyes, being consumed by the universe. Better yet, you can imagine this ghost ship snagged on a reef of stars – the brighter western half of the ship's hull sticking above water, and the fainter part submerged in the shallows. When I really concentrate on the faint eastern extension, the brighter western half flares to prominence in my peripheral vision. It is really a very spooky galaxy.

As the drawing shows, M82 is highly complex and irregular. It takes time and patience to ferret out the details. Most impressive are the energetic-looking bursts of starlight running lengthwise through most of the galaxy. D'Arrest thought M82 scintillated "as if with innumerable brilliant points." At moderate power an obvious elliptical halo surrounds the galaxy's brightest parts. The core is very angular – a sight created by wedgelike lanes of dust. Dark matter prevails on opposing ends of the galaxy. But look closely at the galaxy's extreme western end. Do you see a notch and a kink of stars punctuated with a stellar period? Now, let your eye relax and use keen averted vision. Do you see the very faint "explosions" emanating from the center of the galaxy? Try with high power.

At low power, M81 and M82 share the field of view with two companions, one large and one small. The large one, NGC 3077, is a 10th-magnitude elliptical galaxy 45′ to the southeast of M81, and sports a hurricanelike center. About 1½° southwest of M81 is smaller NGC 2976, a 10th-magnitude spiral seen nearly edge on. Note the little curl on the side away from M81.

In the summer of 1989 I was invited to look at Comet Okazaki–Levy–Rudenko through the 18-inch Clark refractor at Amherst College. The comet floated past M81 and M82 like a specter in a long white gown, and I felt like I was transported into Wilkie Collins's novel *The Woman in White,* just when Walter Hartright first encountered the specter: "I was far too seriously startled by the suddenness with which this extraordinary apparition stood before me, in the dead of night and in that lonely place, to ask what she wanted."

M83

NGC 5236
Type: Spiral Galaxy
Con: Hydra
RA: $13^h 37^m.0$
Dec: –29° 52'
Mag: 7.5
SB: 13.2
Dim: 12'.9 × 11'.5
Dist: 15 million l.y.
Disc: Nicolas-Louis de
Lacaille, 1752

MESSIER: [Observed 17 February 1781] Nebula without a star, close to the head of Centaurus. It appears as a faint, even light, but is so difficult to see with the telescope that the slightest illumination of the micrometer's crosshairs causes it to disappear. It requires considerable concentration to be seen at all. It forms a triangle with two stars estimated to be of sixth and seventh magnitude. Its position has been determined from the stars i, k, and h, in the head of Centaurus. M. de la Caille had already detected this nebula. See the end of this catalogue.

Without question, M83 is my favorite galaxy in the Messier list – a gorgeous face-on spiral. Being in Hawaii naturally enhances the grandeur of this far-southern object, which all too often wallows in atmospheric murk when seen from temperate latitudes. From my observing site M83 is powerfully condensed at low power in the 4-inch, easy in binoculars, and just visible to the naked eye. It is a true showpiece for small telescopes, displaying so much structure – bright knots, dark lanes, and oddly shaped arms – that it *begs* for high power.

Unfortunately, M83 resides in a fairly nondescript part of Hydra, the longest constellation in the sky, so it is not the easiest object to get to. Look for its 7th-magnitude disk 3½° northwest of 1, 2, and 3 Centauri – a tight equilateral triangle of bright stars about 20° (two fist-widths) south of 1st-magnitude Alpha (α) Virginis. If you live under dark skies, use binoculars first to scout out the galaxy before employing your telescope. Adding to the wealth of the telescopic vista is a beautiful string of 10th-magnitude stars bordering the galaxy to the southeast and two fainter stars bracketing its farthest spiral arms. The galaxy immediately looks elongated southeast to northwest, with obvious patches on either side of the nucleus. One patch in the outer, southeast arm is glorious. And with time, several star-forming regions (nebulae) pop in and out of view all over the galaxy. You can spend many enjoyable hours just hunting them down.

To see the most detail, alternate between medium and high power. I use 72 × to see the faintest details in the arms and 130 × to pick out structure in the nuclear region. And I'm amazed at how incredibly detailed the nucleus and its surroundings appear in the 4-inch – how well defined even the tiniest bits of dark matter appear. The nucleus itself is *tiny*, looking like a star trapped in a maelstrom of shimmering light. While concentrating on the nucleus, I notice that a bright knot in the southeastern arm becomes clearly apparent, as does another knot equidistant from the nucleus in the northwestern arm. The knots are oriented in a row that is skewed about 20° from the galaxy's main "bar," which appears to extend across the major axis. The bar is composed of numerous nebulous patches along the galaxy's major axis and ultimately branches off the axis to form the stunning spiral arms. The sight reminds me of the great controversy over Lowell's canals on Mars: in imperfect seeing, the bar looks like a straight line of light, but under excellent conditions, the line breaks up into individual spots.

Once you can follow some of the spiral patterns (and this takes time), try focusing not on the bright arms themselves but on the dark matter between them. The challenge here is to make sense of the innermost region – that maelstrom. I spent hours doing this, to confirm and reconfirm what I was seeing. Take it piecemeal – first look southwest of the nucleus, where the arms appear brighter than those to the northeast. There is one beautiful dark lane, which I think is the most striking feature in the galaxy, just to the northeast of the nucleus. It looks like a black canal whose sandy banks are being illuminated by starlight. On the south side, use averted vision to see a broad arcing band of patchy nebulosity between the nuclear pool and the closest spiral arm. Now spend time concentrating on that arm. When I do this, I discover that the nebulosity separates into two arms; one is more tightly wound than the other and lies closer to the nucleus. You need acute vision and resolution to separate these arms.

The vast and complex network of spiral structure exhibited by M83 could possibly be explained, in part, by tidal warping from a collision with tiny NGC 5253, an elliptical galaxy 10 times less massive than M83 and which now lies about 2° to the southwest.

NGC: A very remarkable object. William and John Herschel found it very bright, very large, elongated in position angle 55°, very suddenly much brighter toward a central nucleus. Seen as a three-branched spiral by Leavenworth [with the 26-inch refractor of Leander McCormick Observatory].

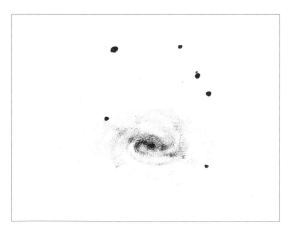

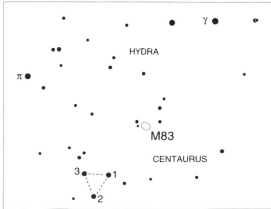

M84

NGC 4374
Type: Elliptical Galaxy
Con: Virgo
RA: 12^h 25^m.1
Dec: + 12° 53'
Mag: 9.1
SB: 12.3
Dim: 6'.5 × 5'.6
Dist: 55 million l.y.
Disc: Messier, 1781

MESSIER: [Observed 18 March 1781] Nebula without a star in Virgo. The center is pretty bright, surrounded by slight nebulosity. Its brightness and appearance are similar to those of numbers 59 and 60 in this catalogue.

NGC: Very bright, pretty large, round, pretty suddenly brighter in the middle, mottled.

M84 and M86 in Virgo seem to have a lot in common: they are separated by a mere 20', they both shine at around 9th magnitude, and they look about the same size. (So, it makes sense to discuss them together, leaving M85 to follow.) But these appearances are deceiving, because M84 and M86 are not the galaxy twins they appear to be.

M84 is an S0 system – a type of galaxy between an elliptical and a spiral – that resides in the Virgo Cloud of galaxies, the nearest of the large extragalactic populations. It stretches across 100,000 light years of space. By contrast, M86 spans 142,000 light years. M86 appears to lie in the heart of the Virgo Cloud. It was once believed to be only a foreground object, residing about halfway between us and the Virgo Cloud, but this is no longer widely accepted.

You can find this curious pair about 5° northwest of Rho (ρ) Virginis. In the 4-inch both exhibit softly glowing disks with condensed centers that gradually give way to diffuse outer haloes. But there are subtle differences. With low power M86 appears slightly larger and brighter than M84 (8.9' in diameter and magnitude 8.9 versus 6'.5 long and magnitude 9.3, respectively). Curiously, my eye is first drawn to M84, whose stellar nucleus and tightly packed core appears more conspicuous than that of M86. Moderate power does little to improve the image. But 130× reveals a 14th-magnitude star southwest of M84's nucleus, which enhances the prominence of the galaxy's core. Do you see another star or condensation equidistant from and northeast of the nucleus?

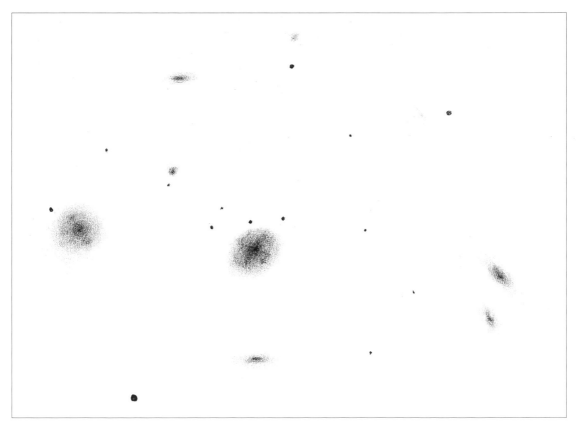

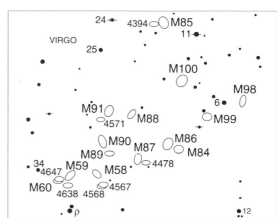

M86

NGC 4406
Type: Elliptical/Spiral Galaxy
Con: Virgo
RA: 12^h 26^m.6
Dec: + 12° 57′
Mag: 8.9
SB: 13.9
Dim: 8′.9 × 5′.8
Dist: 55 million l.y.
Disc: Messier, 1781

MESSIER: [Observed 18 March 1781] Nebula without a star in Virgo, on the same parallel as number 84 above, and very close to it. They have the same appearance, and both appear in the same telescopic field.

NGC: Very bright, large, round, gradually brighter in the middle to a nucleus, mottled.

Now use high power to concentrate on M86. It too has a sharp nucleus, but its core is more diffuse than the core of M84. I see a very curious feature in this nuclear haze, which can only be described as a swastika – the ancient cosmic symbol of good luck (but of tragedy today) – glazing an otherwise porcelain-pure oval.

A minefield of galaxies, the Virgo Cluster must have posed quite a problem for Messier and his contemporaries as they swept this region for comets. So many nebulous glows went unrecorded by these skilled observers that I wonder if, in trying to plot them, utter confusion was the culprit more than the optical limitations of their telescopes! For the patient observer under a dark sky, the region centered on M84 and M86 just might be one of the richest galaxy fields visible in small telescopes. Although few of the objects display impressive details (most look like fuzzy patches of varying brightnesses and sizes), they nevertheless have an enchanting aura about them. What the Virgo Cloud cannot offer in quality, it makes up for in quantity. Fully 10 galaxies fill a 1° field centered on M84 and M86. And the entire Virgo Cloud – which is more properly referred to as the Coma–Virgo Cloud, because the greatest concentration of galaxies in the region huddle close to the border of these two constellations – contains at least 10 times that many galaxies within reach of a 4-inch under dark skies. But not all are easy to see. It takes a dedicated effort, for example, to sight each individual object shown in the accompanying drawings.

M85

NGC 4382
Type: Lenticular Galaxy
Con: Coma Berenices
RA: 12^h 25^m.4
Dec: + 18° 11′
Mag: 9.1
SB: 13.0
Dim: 7′.1 × 5′.5
Dist: 55 million l.y.
Disc: Pierre Méchain, 1781

MESSIER: [Observed 18 March 1781] Nebula without a star, above and close to the ear of corn in Virgo, between two stars in Coma Berenices, Flamsteed 11 and 14. This nebula is very faint. M. Méchain determined its position on 4 March 1781.

NGC: Very bright, pretty large, round, with a bright, middle star north preceding [west].

Although it resides in Coma Berenices, M85 belongs to the Virgo Cloud of galaxies. Ironically, this lenticular system is the real twin to M84 in Virgo. M84 and M85 both are about 55 million light years distant, both shine at 9th magnitude, and they have masses of 500 billion and 400 billion suns, respectively. And like M84, M85 is a gas-poor galaxy that displays low-contrast spiral structure.

M85 is easily spotted a little more than 1° east of the 4.7-magnitude star 11 Comae Berenices and 5° north of M84. Through the 4-inch at 23× the galaxy appears as a powerfully glowing oval mass with a distinct nucleus. It shares a high-power field of view with two 10th-magnitude stars and the 11th-magnitude barred-spiral NGC 4394 to the northeast. M85 forms a strongly interacting pair with NGC 4394. See if you notice a slight blue coloration to M85.

The telescopic view at high power is reminiscent of photographs I have seen of this pair taken with large instruments! A glorious 12.5-magnitude star blazes just to the north-northeast of the galaxy's nucleus. An arc of bright material to the south, which seems to have escaped the gaze of many visual observers, caught me by surprise. The only reference I have found that refers to this peculiar yet distinct feature is *Burnham's Celestial Handbook*, where Burnham says plates taken at Palomar show "faint elongations or tufts of material at the north and south tips of the system, either the vague beginning of a spiral pattern or the last surviving remnant of one." Do you also see a bar running through the galaxy's major axis?

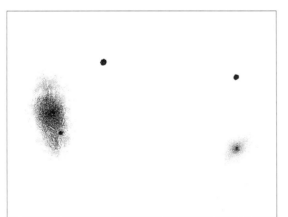
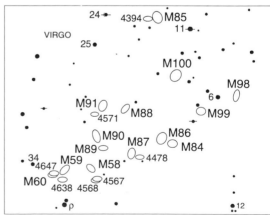

M87

NGC 4486
Type: Elliptical Galaxy
Con: Virgo
RA: 12^h 30^m.8
Dec: + 12° 24′
Mag: 8.6
SB: 12.7
Dim: 8′.3 × 6′.6
Dist: 55 million l.y.
Disc: Messier, 1781

MESSIER: [Observed 18 March 1781] Nebula without a star in Virgo, below and quite close to an eighth-magnitude star. The star has the same right ascension as the nebula, and its declination is 13° 42′ 21″ north. This nebula appears to have the same brightness as the two nebulae numbers 84 and 86.

NGC: Very bright and large, round, much brighter in the middle, third of three [easternmost].

A monstrous ball of energy, M87 has a total mass of nearly 800 *billion* suns, making it one of the more massive galaxies known. It is also one of the brighter ellipticals in the Virgo Cluster and one of the more luminous of all known ellipticals. M87 is easily located by drawing a line from M58 to M84; M87 lies almost equidistant between them, being slightly closer to M84 and just 6′ south of an 8th-magnitude star.

At 23× the 8.6-magnitude elliptical can be seen glowing to the east of a fine telescopic asterism that looks a bit like Leo, the Lion. The galaxy itself resembles an unresolved globular cluster, one with a bright spherical shell that gradually condenses inward. I cannot see a sharp nucleus, nor much other detail, even at high power.

But don't let its wimpy appearance fool you. M87 is a cosmic powerhouse with a peculiar jet of matter blasting out from its nuclear region. After studying Hubble Space Telescope images of M87, astronomers now believe that a black hole with a mass of 3 billion suns lurks in the galaxy's nucleus, within a zone about 120 light years across. High-resolution terrestrial images reveal a beaded jet of hot ionized gas, called plasma, though the brightest beads are barely within the visual range of large-aperture amateur instruments. Barbara Wilson and I spied evidence of the jet in her 20-inch reflector during a Texas Star Party. The jet extends nearly 5,000 light years from M87 and points toward M84. It was once believed to be an intergalactic bridge between them, but the mysterious feature probably has a more interesting origin – the voracious black hole lurking at the center of M87, whose nuclear region is swarming with billions of rapidly orbiting suns. The black hole is already thought to have ingested clouds of gas (about 5 billion solar masses worth) and some whole stars, as well. If you consider that M87's jets are only about as massive as our sun,

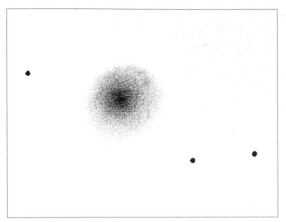

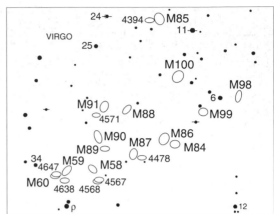

they may be the ejected remains of a star that was sucked into the black hole.

Don't miss tiny NGC 4478, an 11th-magnitude elliptical 10′ south-west of M87. At 23× the galaxy looks merely like a star, but higher magnification brings it clearly into view. I also spotted a close double star to the west, which resembles the light of a galaxy.

M88

NGC 4501
Type: Spiral Galaxy
Con: Coma Berenices
RA: 12^h 32^m.0
Dec: + 14° 25′
Mag: 9.6
SB: 12.6
Dim: 6′.9 × 3′.7
Dist: 55 million l.y.
Disc: Messier, 1781

MESSIER: [Observed 18 March 1781] Nebula without a star in Virgo, between two faint stars and a sixth-magnitude star, which all appear in the same telescopic field as the nebula. Its luminosity is one of the faintest and it resembles that reported for number 58, also in Virgo.

NGC: Bright, very large, very much extended.

John Mallas aptly described M88 as resembling the Andromeda Galaxy (M31) depicted in a very small-scale photograph. Lying about 4½° southeast of 11 Comae Berenices, M88 is one of the more conspicuous spiral systems in the Virgo Cluster of galaxies. Although it shines at only magnitude 9.5, its compact size and 30° tilt from edge on give it good surface brightness.

At 23×, the galaxy displays a definite spindle shape and is easily picked out from the galactic morass. No other galaxies near it shine as brightly, and a close pair of 12th-magnitude stars (separated by 30″) touch the southeast extremity of the nebulous glow. An 11th-magnitude star is visible to the north.

Moderate power reveals a strong suggestion of spiral structure along the western and eastern rims. But it took me a few nights and many hours to plot the fainter arms closer to the nucleus. Although I could detect

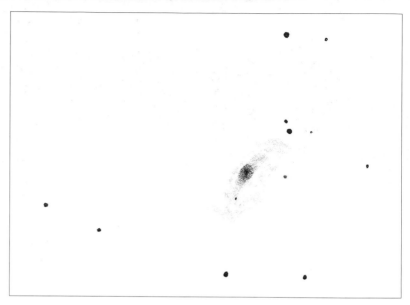

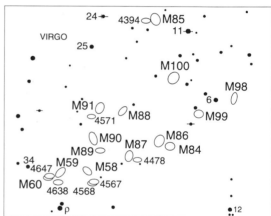

them – and, in moments of excellent seeing, they appeared obvious – I had difficulty translating those impressions to paper. The problem is that most of the brighter inner detail shines with a uniform glow that is broken in places by dark matter.

The one interesting feature of this galaxy, however, seems to be a mystery. When I made my first drawing of M88, it showed the galaxy's pin-point nucleus and outer spiral arms. I noted the galaxy's position between the three foreground stars mentioned previously. In a drawing I made 10 nights later, however, I included a 13.5-magnitude star southeast of the nucleus (indicated with a question mark in the drawing), which I had apparently not seen the first time. Moreover, the photographs I have seen of M88, including the accompanying one, do not show any star in or near that position. Was this "new" star a supernova? An asteroid passing by? Or is it a foreground variable star? Give the galaxy a look and see what you think.

M89

NGC 4552
Type: Elliptical Galaxy
Con: Virgo
RA: 12^h 35^m.7
Dec: + 12° 33'
Mag: 9.7
SB: 12.3
Dim: 5'.1 × 4'.7
Dist: 55 million l.y.
Disc: Messier, 1781

MESSIER: [Observed 18 March 1781] Nebula without a star in Virgo, not far from, and on the same parallel as the nebula, number 87, described above. Its light is extremely faint and diffuse, and it is visible only with difficulty.

NGC: Pretty bright and small, round, gradually much brighter toward the middle.

The Virgo Cluster harbors some 250 large galaxies and perhaps a thousand or more smaller systems. One distinguishing characteristic of this particular galaxy cluster is its smattering of spirals loosely dispersed among the more amorphous ellipticals. This peculiarity is nicely represented by the pairing of M89 and M90, an elliptical and a spiral, respectively. Separated by only 40', these galaxies should fit in the low-power field of almost any amateur instrument. Although their magnitudes are similar, M90 is bigger, bolder, and more interesting than its counterpart.

With a diameter of 150,000 light years and a mass of 80 billion suns, M90 is one of the larger spirals in the Virgo Cluster. It forms the northern apex of an equilateral triangle with the brighter galaxies M58 and M87. To help you further identify the pair, a line from M90 through M89 would bisect the equilateral triangle. There is also a zigzagging chain of faint stars connecting the two galaxies.

At 23× M89's slightly swollen starlike disk is easy to overlook, especially with mighty M90 to its northeast. Yet I find the elliptical strikingly beautiful at 72×, its silvery light pure and pleasing to the eye. The galaxy does have an extremely sharp nucleus surrounded by a misty glow; high

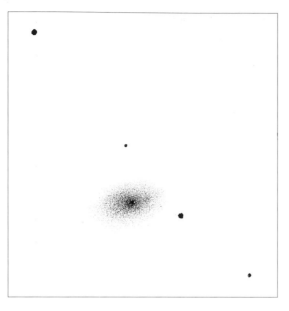

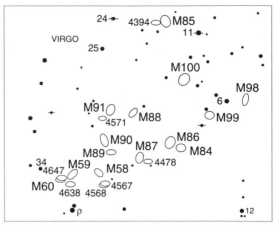

M90

NGC 4569
Type: Spiral Galaxy
Con: Virgo
RA: 12^h 36^m.8
Dec: + 13° 10′
Mag: 9.5
SB: 13.6
Dim: 9′.5 × 4′.4
Dist: 55 million l.y.
Disc: Messier, 1781

MESSIER: [Observed 18 March 1781] Nebula without a star in Virgo. Its light is as faint as the previous one, number 89.

NGC: Pretty large, brighter in the middle toward a nucleus.

power brings out suggestions of irregularities in the halo. For instance, there seems to be, though I am not certain of this, a knot to the east, which makes that side of the galaxy look brighter than the other. That knot might be a very faint star.

M90, on the other hand, is a strong oval glow – a dimly shaped spiral whose major axis is elongated slightly southwest to northeast. But close

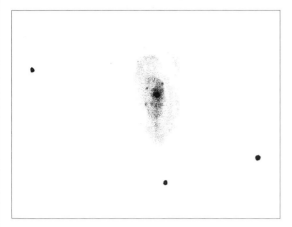

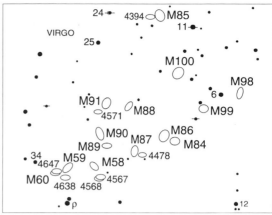

inspection shows a bar of light across the nucleus with a north–south orientation. Clawlike arcs of light curve off to the west from each end of the nuclear region, giving it the shape of a crab. The outer spiral arms make a slightly warped S-shaped pattern. Skiff noticed that the western flank of the halo bulges in the center and looks flattened to the east. The nuclear region also looks mottled, either with faint starlight or dark nebulosity.

M91

Missing Messier Object
NGC 4548
Type: Barred Spiral Galaxy
Con: Coma Berenices
RA: 12^h 35^m.4
Dec: + 14° 30′
Mag: 10.1
SB: 13.3
Dim: 5′.4 × 4′.3
Dist: 55 million l.y.
Disc: Johan Elert Bode, 1774

"Of Messier's 14 nebulae in [the Coma–Virgo] region, 13 are easily identified today, but M91 cannot be found. . . . There is simply no bright nebula omitted by Messier that could conceivably be identified with M91." Thus wrote a perplexed Owen Gingerich in a chapter he contributed to Mallas and Kreimer's *Messier Album*. Gingerich opened a Pandora's box of possibilities. Was M91 a comet, as Shapley and many others had proposed? Is it NGC 4571, the closet galaxy to Messier's recorded position, or a duplicate observation of M58 (a simple mistake made by Messier), as Gingerich suggests? Perhaps it is just as d'Arrest supposed: "The nebulae M91 in this position, no longer exists in the heavens."

The most widely accepted explanation today is that Messier's 91st catalogue entry was actually NGC 4548, a 10th-magnitude barred spiral galaxy. So argued W. C. Williams of Fort Worth, Texas, in the December 1969 issue of *Sky & Telescope*. He described how Messier probably made a mistake in both reduction and plotting. By applying a mathematical correction to Messier's right ascension and declination for the object, Williams encountered NGC 4548, about 4½° northwest of 4.9-magnitude Rho (ρ) Virginis, and deduced that this was the missing Messier object, M91.

When I use low power to look at the field containing M91 I am immediately impressed by how obvious this galaxy appears compared to NGC 4571 – a smaller, 11th-magnitude spiral ½° to the southeast, which would certainly have been beyond Messier's visual grasp. NGC 4571 is faint, even under superb Hawaiian skies through superb optics! Also, would not Messier have noted the bright star 30′ to its southeast?

At 23× M91 is a very modest glow, definitely out of round – a broad spiral shape with a bright center. Medium power begins to display its

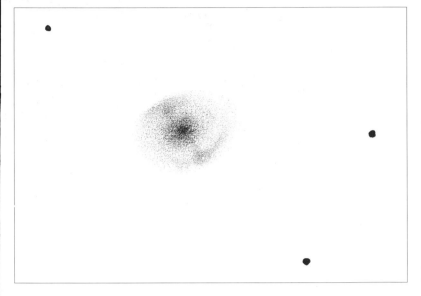

MESSIER: [Observed 18 March 1781] Nebula without a star in Virgo, above the preceding one, number 90. Its light is even fainter than that one.

Note: The constellation of Virgo, and above all the northern wing, is one of the constellations that contains the most nebulae. This catalogue includes 13 that have been detected, namely numbers 49, 58, 59, 60, 61, 84, 85, 86, 87, 88, 89, 90, and 91. All these nebulae appear to be without stars. They can be seen only under extremely good skies, and close to meridian passage. Most of these nebulae have been pointed out to me by M. Méchain.

NGC: A remarkable object, extremely bright and extremely large, extended in position angle 156°. It increases in brightness inward, first gradually and then suddenly to a very much brighter center. Bright nucleus.

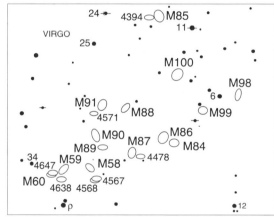

complexities, such as a central bar, knots, and hazy arcs outlining the apparent halo. High power, and persistence, are required to trace out the arms (I had to really breathe hard and use averted vision). Interestingly, Jones thought a 12-inch was required to see the central bar! But look at my pencil rendering. Perhaps I really did not see the physical bar in my 4-inch; maybe my eye was merely connecting the dots – the starlike nucleus and the knots at the ends of the bar. Study it with your scope and see what you think.

M92

NGC 6341
Type: Globular Cluster
Con: Hercules
RA: 17^h 17^m.1
Dec: +43° 08'
Mag: 6.5
Dia: 14'
Dist: 25,400 l.y.
Disc: Johan Elert Bode, 1777

MESSIER: [Observed 18 March 1781] A fine, conspicuous nebula, very bright, between the knee and left leg of Hercules. It is clearly visible in a one-foot refractor. It does not contain any stars. The center is clear and bright, surrounded by nebulosity and resembles the nucleus of a large comet. Its light and size are very similar to those of the nebula in the belt of Hercules. See number 13 in this catalogue. Its position was determined by comparing it directly with the fourth-magnitude star σ Herculis. The nebula and the star lie on the same parallel.

NGC: Globular cluster of stars, very bright, very large, extremely compressed in the middle, well resolved, small [faint] stars.

Just as globular cluster M28 in Sagittarius is commonly overlooked in favor of the more dazzling globular M22 nearby, so M92 in Hercules plays second fiddle to the much-vaunted M13. As Burnham notes, if M92 were in any other constellation, it would be considered a showpiece. Located 5° southwest of Iota (ι) Herculis, this 6.4-magnitude bolus is a surprisingly easy naked-eye catch, even though it is about a magnitude fainter than M13. One night I discovered that the two Hercules globulars can be seen simultaneously with the unaided eye! Try this from your observing site, and see if you don't surprise yourself. Both globulars lie about the same distance from us, so you are really seeing a difference in physical size, reflected in the objects' brightnesses. M92 is roughly 10 light years smaller

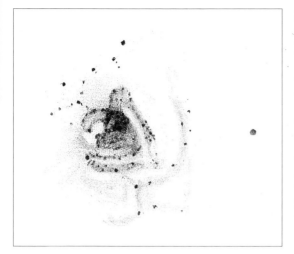

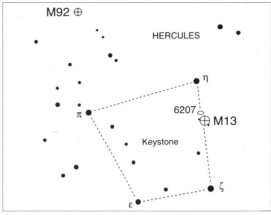

in actual diameter than M13, but in angular diameter M92 spans 14′ compared to M13's 21′.

At 23×, M92 displays a very tight core and some definition in the outer halo; that is, it looks faintly granular and elongated north to south. But overall the globular appears much more compact than M13. Resolution begins at 72×, while 130× and averted vision transform the cluster into myriad points of light that appear to mushroom from a very tight core. The region immediately surrounding the core is not as uniform as it might first appear. In fact, I find it strongly asymmetrical to the north and shaped somewhat like a lobster claw. Close examination of the void between the claw's pincers reveals a very faint arc of stars enclosing the darkness to the west. So the void now looks like a keyhole. Streams of stars flow out of the core and nearly wrap themselves around the cluster's northeastern part, which looks rather heart shaped. These streamers remind me of the wings of stars flowing from the core of M13, but in M92 they are flowing in the opposite direction.

Not everyone will discern all of the features I have described here. The stars in M92 vary so much in brightness and are in such complex arrangements that I can only liken the view to a Rorschach test, the inkblot test used in psychology: at one moment I see a face-on spiral galaxy (Rosse, by the way, saw a spiral contour to the nucleus), at another, I see two rings of starlight perpendicular to each other. I blink and the lobster claw materializes.

Finally, to add to the confusion, M92 is riddled with dark lanes. In fact, when Isaac Roberts looked at his photographs of M92 taken through a 20-inch telescope, he believed the cluster to be involved with dense dark matter that almost prevented stars from being seen through it. Aside from the keyhole, I believe I have found one other obvious dark pattern, a wishbone, in the northeast part of the cluster.

M93

NGC 2447
Type: Open Cluster
Con: Puppis
RA: 7^h 44^m.5
Dec: −23° 51′.2
Mag: 6.2
Dia: 10′
Dist: 3,600 l.y.
Disc: Messier, 1781

MESSIER: [Observed 20 March 1781] Cluster of faint stars, without nebulosity, between Canis Major and the prow of Argo Navis.

NGC: Cluster, large, pretty rich, little compressed, with 8th- to 13th-magnitude stars.

There are three reasons why I consider this open cluster in Puppis a stunning visual treat. First, its stars are like tiny gems scattered in a rich field of even tinier gems, these are subtle beauties, as delicate as pearls on white satin. Second, the core of the cluster has a distinct arrowhead shape, much like that of M48 in Hydra. And third, in binoculars, M93 appears asymmetrical, and I can see a "cat's eye" pattern, similar to that in M4.

Perched fairly low in the sky from temperate latitudes, M93 sits about 1½° northwest of Xi (ξ) Puppis, a 3rd-magnitude yellow supergiant star 7° east-northeast of Eta (η) Canis Majoris, in the tail of the Great Dog. Still, Walter Scott Houston said he could detect it from his home in New England without optical aid. Telescopically, the cluster fires the imagination. Smyth fancied a starfish pattern to the stars. (Smyth also shared a somewhat amusing story of how Chevallier d'Angos of the Grand Master's Observatory in Malta once mistook M93 for a comet, which caused Baron de Zach to call such astronomical blunders "Angosiades.") Jones visualized a butterfly with open wings, and Skiff saw three rows of stars emanating from the center like "trident spikes." To me, M93's V shape is part of a sunlit spider resting in the center of its dew-laden web, as you can see from my sketch.

M93 contains about 80 stars in a 10′ field. Do you get the impression that the two brightest stars in the southwest portion of the arrowhead appear yellowish orange? I see this only at 23×. Be sure to look all around the arrowhead, because a swarm of fainter stars appears to form a glittering spherical halo around it. The sight is almost mystical.

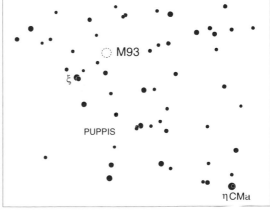

M94

Croc's Eye Galaxy
NGC 4736
Type: Spiral Galaxy
Con: Canes Venatici
RA: 12^h 50^m.9
Dec: +41° 07'
Mag: 8.2
Dim: 11'.2×9'.1
Dist: 14 million l.y.
Disc: Pierre Méchain, 1781

MESSIER: [Observed 24 March 1781] Nebula without a star, above Cor Caroli [α Canum Venaticorum], on the same parallel as the sixth-magnitude star Flamsteed 8 Canum Venaticorum. The center is bright and the nebulosity not very diffuse. It resembles the nebula that lies below Lepus, number 79, but this one is finer and brighter. M. Méchain discovered it on 22 March 1781.

NGC: Very bright, large, irregularly round; very suddenly much brighter toward the middle to a bright nucleus; mottled.

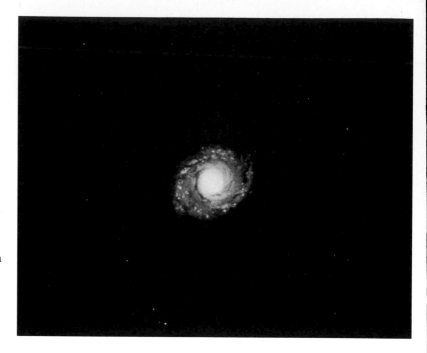

Despite its placid appearance, M94 may have experienced a violent explosion only 10 million years ago – a cleansing event that disgorged millions of solar masses of material out of its disk or nucleus. Episodes of violent behavior may be typical for spiral systems; if so, we are seeing this mundane-looking spiral in a quiescent state. Its disk is oriented nearly face on to us, and photographs through large instruments reveal its tightly wrapped arms, reminiscent of those in satellite views of a tropical hurricane.

This compact galaxy looms brightly about 3° northwest of the double star Alpha[1,2] ($α^{1,2}$) Canum Venaticorum, or 1½° northeast of the midpoint between Alpha and Beta (β) Canum Venaticorum. At 23×, the galaxy can be seen in the neck of a Cygnus- or Grus-like asterism, shining with an almost starlike radiance, though a careful look will reveal it to be surrounded by a much larger, albeit fainter, glow.

The galaxy is a most intriguing sight at moderate to high power, because, with time, it seems to peer right back at you, like a hypnotic eye. A near opposite of the famous Black Eye Galaxy (M64), whose central region is lined with dark material, M94 sports a faintly luminous ring that surrounds a sharp nucleus. This ring reminds me of the yellow circle around a crocodile's eye – a discriminating, all-the-better-to-see-you-with marking. (Can you make out the southern curl of this ring?) The crocodile metaphor is also fitting because, as described earlier, this is a "violent" galaxy.

The 48-inch Schmidt camera on Palomar Mountain recorded another faint elliptical ring outside M94, perhaps debris blown out from the bellowing episode. It measures about 15′ in diameter (M94 is only 11′.2 in diameter on its major axis), has a sharp inner edge, and reveals a major axis skewed about 30° from that of M94. One would think that this ring would be out of reach of amateur instruments, so why bother? Well, when I swept the 4-inch back and forth across that field several times at 23×, I convinced myself that this outer ring might be at the visual threshold. My drawing of its location, extent, and orientation, appears to be consistent with the one in the 48-inch image. Am I simply being misled by the power of suggestion? Or is this ring real and within reach of amateur instruments? Many times an object discovered with large telescopes turns out to be visible in small telescopes once knowledge of it exists. I'll let you be the judge.

Speaking of possible illusions created by the power of suggestion, d'Arrest thought M94's circular disk and central star were similar to those of a planetary nebula. He even saw a "sky-blue" color!

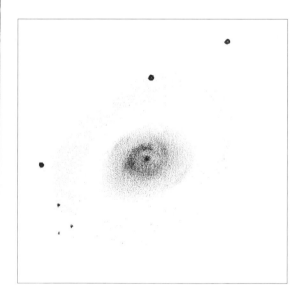

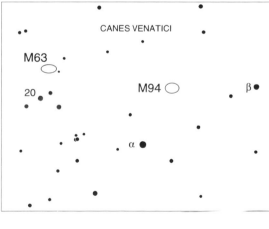

M95

NGC 3351
Type: Barred Spiral Galaxy
Con: Leo
RA: $10^h 44^m.0$
Dec: $+11° 42'$
Mag: 9.7
SB: 13.5
Dim: $7'.4 \times 5'.0$
Dist: 26.5 million l.y.
Disc: Pierre Méchain, 1781

MESSIER: [Observed 24 March 1781] Nebula without a star in Leo, above the star *l* [53 Leonis]. Its light is very faint.

NGC: Bright, large, round, gradually much brighter toward a nucleus.

M95 and M96, two spiral systems in Leo, fit comfortably together in the field of view of 7×50 binoculars, separated by only 42'. Huddling with them in the same field is the elliptical galaxy M105. They lie 8° west of another Leo galaxy trio, consisting of M65, M66, and NGC 3628. All belong to a small cluster of galaxies known as the Leo Galaxy Group, whose members are speeding away from us at 800 miles per second.

M96 has been the focus of cosmologists' attention of late. Astronomers used the Hubble Space Telescope (with its improved optics) to probe the galaxy's spiral arms in search of Cepheid variable stars. The period of a Cepheid's variability is strictly related to the intrinsic luminosity of the star. By comparing how bright a Cepheid looks with how luminous it really is, astronomers can determine the star's distance. Using this stellar yardstick on Cepheids in M96 and comparing them with those in the nearby Large Magellanic Cloud, astronomers determined that M96 lies 38 million light years away – nearly 60 percent farther away than previously thought. (For consistency, however, the distances used in the data lists in this book were derived from the sources indicated in chapter 3.) This implies that *all* the galaxies in the Leo Group might be that much more distant than earlier calculations had put them.

You can find M95 and M96 lurking about 9° east of Regulus (Alpha [α] Leonis) and 1½° northwest of 5th-magnitude 53 Leonis. Although I did not estimate their magnitudes, my gut feeling is that both are brighter than the values listed for them (9.7 for M95, and 9.2 for M96), perhaps by a half magnitude or more. At low power M95 appears larger and more diffuse than its neighbor M96, whose core is much tighter.

M95 is a remarkable object that reminds me of Darth Vader's Death Ship in *Star Wars*, because it has a round center with bars extending out to

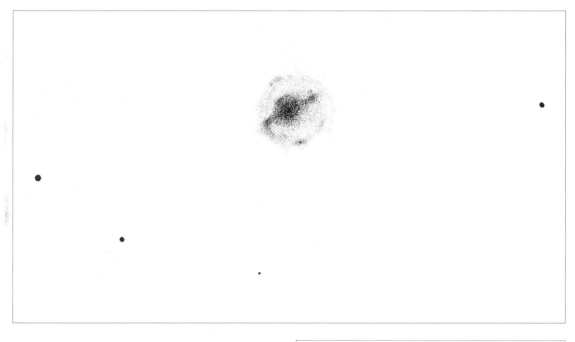

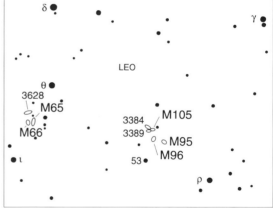

arcing "wings." It shows hints of irregularity even at a glance, but high power is needed to resolve the fine details hiding in the inner region. At 130× the central bar shows up with little difficulty, and it seems to have knots at each end. (Mallas, using his 4-inch f/15 refractor, could not see the bar.) With its central disk, narrow arms, and beads of light, the galaxy looks a little like an image of Saturn with near edge-on rings. But this illusion vanishes with a longer look. The bar stretches away from the center and connects to a faintly luminous, slightly beaded ring that completely encircles the bar. Now the galaxy looks like an aerial view of an island

M96

NGC 3368
Type: Spiral Galaxy
Con: Leo
RA: 10^h 46^m.8
Dec: +11° 49′
Mag: 9.2
SB: 12.9
Dim: 7′.6×5′.2
Dist: 26.5 million l.y.

Disc: Pierre Méchain, 1781

MESSIER: [Observed 24 March 1781] Nebula without a star in Leo, near the preceding one [M95]. This one is less conspicuous. Both are on the same parallel as Regulus. They resemble the nebulae in Virgo, numbers 84 and 86. M. Méchain saw both on 20 March 1781.

NGC: Very bright and large, little extended, very suddenly much brighter in the middle, mottled.

covered by clouds and surrounded by a coral atoll. An impressive knot adorns the northern part of the ring.

The core of M96 is much more condensed than that of M95. A tiny nucleus lies within the strong central glow, but high power is needed to bring it out well. After studying the galaxy's faint textures at 130×, I noticed a detached arc of light just southwest of the elliptical nucleus. It appears to be the brightest part of a separate arm that branches off the southeastern side of a warped central bar; the southeastern tip of this bar curls westward, while the northwestern tip curls eastward. So the galaxy looks like a barred spiral that is on the verge of becoming fully spiral. See if you don't start to feel uncomfortable the longer you look at M96; I find its shape to resemble that of an eye – a piercing, almost sinister one at that!

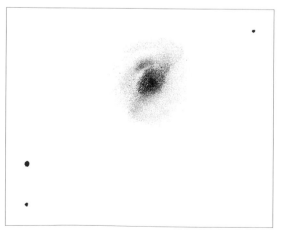

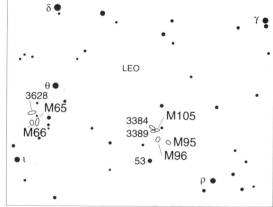

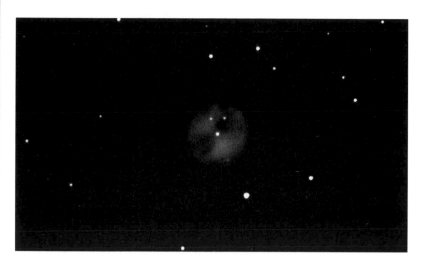

M97

Owl Nebula
NGC 3587
Type: Planetary Nebula
Con: Ursa Major
RA: 11^h 14^m.8
Dec: + 55° 01′
Mag: 9.9
Dia: 170″
Dist: 1,630 l.y.
Disc: Pierre Méchain, 1781

MESSIER: [Observed 24 March 1781] Nebula in Ursa Major, near β. M. Méchain reports that it is difficult to see, especially when the micrometer crosshairs are illuminated. Its light is faint, and it is without any stars. M. Méchain first saw it on 16 February 1781, and the position given is the one that he reported. Close to this nebula he saw another, whose position has not yet been determined, as well as a third, which is close to γ Ursae Majoris.

NGC: Very remarkable planetary nebula, very bright and large, round, 150 seconds in diameter. It brightens toward the middle very gradually, then suddenly.

This distinctive planetary nebula, so strongly reminiscent of the face of an owl (at least in photographs), is a real challenge for small telescopes because its faint light is spread over an area roughly five times the size of Jupiter's apparent disk. Thus, to see the Owl Nebula well, one needs to be under a dark sky. In *Messier's Nebulae and Star Clusters*, Jones expounded on the difficulty of seeing this object. He noted that the glow will probably elude anyone using a telescope with an aperture smaller than 6 inches and that one needs at least a 12-inch instrument to see the "eyes" of the Owl. In that work the nebula is listed as 12th magnitude, though Jones said it is slightly brighter.

But I have to disagree with Jones about the Owl's elusiveness. M97 is one of the finer examples of a planetary nebula for a 4-inch telescope, displaying a pale but intricate shell of nebulosity. You have to keep a fixed gaze, though, to see its delicate details. It earned the "Owl" moniker from a sketch made by Rosse in 1848, who through his 72-inch telescope spied the two dark holes that look like eyes peering spookily from a round face. The Owl is certainly challenging, especially from a city, but I can recall seeing it many years ago from Cambridge, Massachusetts, with a 4-inch reflector. And Mallas definitely saw the Owl's eyes with a 4-inch refractor in the dark skies over Arizona. From Hawaii, M97 is *just* visible in 7 × 35 binoculars. The field is easily located 2° southeast of Merak (Beta [β] Ursae Majoris), a blue A1 main-sequence star in the bowl of the Big Dipper. In binoculars the Owl forms the eastern arm of an asterism resembling the Southern Cross (Crux). Through the Genesis at 23 × the nebula is a strong ball of light surrounded by a thinner, weaker shell, which is slightly off-center – reminiscent of a composite print out of registration.

To see the innermost detail you have to experiment with different eyepieces and determine which one gives you the highest magnification without adversely diminishing the object's brightness. Jumping back and forth from 72× to 130× works best for me. At medium power the glow from the Owl promptly appears as a nonuniform disk. Every now and then the 14th-magnitude central star pops into view, but I have seen it only occasionally with averted vision. It may be the single most challenging feature in the entire nebula. (If you do see it, take a moment to consider that the entire nebula is lit by the light from this white-dwarf star about the size of earth.)

The Owl's eyes can be essentially inferred at 72× by sweeping your gaze back and forth across the disk. When I do this I see a small smoke ring, like the Ring Nebula. At high power two stars appear on the southern side, and I catch hints of the central star. The northern side of the planetary is brighter, slightly off-axis, and strongest toward the northeast. The eyes themselves are rather difficult to resolve, and it takes a lot of time behind the eyepiece to figure out their orientation. Rather than trying to see dark holes, try instead for the dim bar of light that separates them. Don't be fooled, however, by the brighter northern outline of the eye sockets, which can trick you into believing that *it* is the bar. This pseudo-bar might be responsible for an optical illusion, which I call the "black-lash phenomenon." With low power and strong peripheral vision, a mysterious curved lash of darkness arcs across the glowing disk and wavers in and out of view. When I change the tilt of my head, the axis of

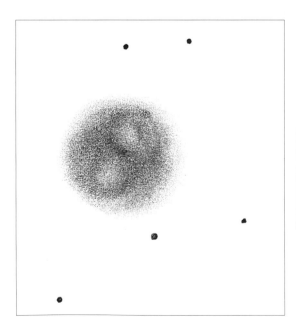

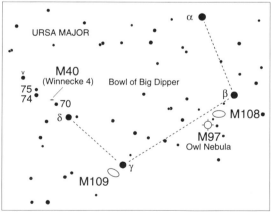

this phantom lash also changes. I have also seen it flip-flop between two different positions. By the way, I have not read an account of anyone seeing color (outside of gray or silver) in this planetary. What do reflector owners have to say about this?

Coincidentally, when I finished observing M97 on the third and final night, I looked skyward and heard the hoot of a Pueo, the Hawaiian short-eared owl.

M98

NGC 4192
Type: Spiral Galaxy
Con: Coma Berenices
RA: 12^h 13^m.8
Dec: + 14° 54′
Mag: 10.1
SB: 13.2
Dim: 9′.8×2′.8
Dist: 55 million l.y.
Disc: Pierre Méchain, 1781

MESSIER: [Observed 13 April 1781] Nebula without a star, whose luminosity is extremely faint, above the northern wing of Virgo, and on the same parallel as, and close to the fifth-magnitude star, Flamsteed 6 in Coma Berenices. M. Méchain saw it on 15 March 1781.

NGC: Bright, very large, extended along position angle 152°, suddenly very much brighter toward the middle.

What a refreshing sight! The nearly edge-on galaxy M98 in Coma Berenices offers the small-telescope user more detail to hunt down than most of the face-on spiral and elliptical systems in this region. Its pale, slender disk, hooklike extensions, and dark dust make it a soothing yet enticing sight. Like M86, this 130-billion-solar-mass metropolis lies in the great Coma–Virgo galaxy cloud. But unlike most of the other members of the cloud, which are receding from us at several hundred miles per second, peculiar M98 appears to be *approaching* us at 142 miles per second.

The galaxy, a 10th-magnitude spiral, can be found just ½° east of 5th-magnitude 6 Comae Berenices, so you will have to use high magnification to exclude that star from the field of view. Through the 4-inch the most obvious feature of the galaxy, aside from its highly elongated shape, is a fan of material southwest of the nucleus, which when studied more closely seems composed of bright and dark patches reminiscent of those in M108. The inner nuclear region has spiral structure that abruptly terminates along the eastern rim, yielding semicircular arms that look like pincers – much more prominent and dramatic than the crablike claws of M90. These features of M98, created by a remarkable dustiness, are best seen with high power. Now switch to a low-power eyepiece. Do you see how the entire galaxy is surrounded by an elongated halo that stands out boldly against the dark background? For *Star Trek* fans, does this galaxy remind you of a Klingon vessel?

Admittedly, the first night I happened to view M98 turned out to be one of the clearest nights in this observing project. That evening, the Milky Way stood above the horizon like a god. It struck me as ironic that, after spending so many hours squinting at dim and distant M98, I could look away from the eyepiece to see *our* galaxy spread so boldly across the sky that I thought it would crush me.

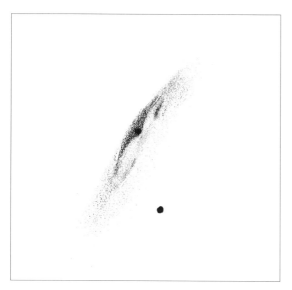

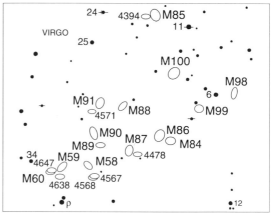

M99

NGC 4254
Type: Spiral Galaxy
Con: Coma Berenices
RA: 12^h 18^m.8
Dec: + 14° 25′
Mag: 9.9
SB: 13.0
Dim: 5′.4 × 4′.7
Dist: 55 million l.y.
Disc: Pierre Méchain, 1781

MESSIER: [Observed 13 April 1781] Nebula without a star, whose luminosity is very dim, but which is slightly clearer than the previous one [M98], lying on the northern wing of Virgo, and close to the same star, Flamsteed 6 in Coma Berenices. The nebula is between two stars of the seventh and eighth magnitude. M. Méchain saw it on 15 March 1781.

NGC: Very remarkable. William and John Herschel called it bright, large, round, gradually brighter in the middle, mottled. F. P. Leavenworth and Rosse saw it as a three-branched spiral.

Just 1½° southeast of the nearly edge-on spiral M98 in Coma Berenices looms the face-on pinwheel M99. The two galaxies present a nice morphological contrast together in the low-power field of view. Another way to locate M99 is to look 50′ southwest of the 5.1-magnitude star 6 Comae Berenices, 2° northeast of which beckons M100.

In photographs made with large-aperture instruments M99 shows two thin spiral arms that fly off to the west from the main central mass and one strong arm that arches far to the east. Several minor tufts of starlight also stray off the nuclear region to the west. The telescopic image in the 4-inch is but a pale ghost of that mighty spectacle. Yet details do emerge with effort.

At 23× a beadlike nucleus is set inside a round, unassuming outer glow. Increasing to moderate magnification transforms M99 into a subtle beauty. The two strongest arms materialize slowly, soon displaying graceful knots that look like diamonds balanced on a fingertip. This is a 9.8-magnitude galaxy with an angular size of 5′.4 × 4′.7 and a low overall surface brightness. The quality of your skies may well determine whether or not you see any spiral structure at all. Roger Clark in his book, *Visual Astronomy of the Deep Sky*, reported seeing "no hint" of spiral detail in his 8-inch Cassegrain under "moderate to good" sky conditions.

Astronomically, M100 is one of the larger spiral systems in the Coma–Virgo Cloud of Galaxies, comparable in size to our own galaxy. Through a telescope, however, M100 has little to offer. Although at magnitude 9.4 it is brighter than M99, it is nevertheless a more difficult object because its light is spread over an even larger surface area (7′.4 × 6′.3). It remains a pale orb even with high magnification, though perseverance will show traces of ghostly arms. Photographically, M100 is supremely tex-

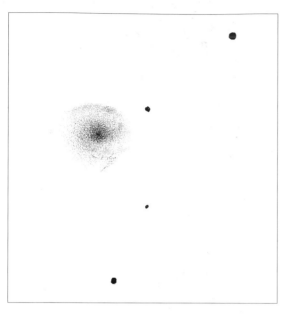

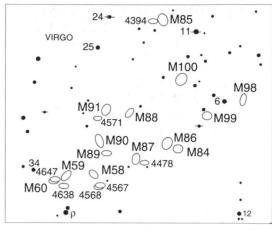

tured, its spiral appendages interspersed with small patches of obscuring matter. The moral here is not to become too spoiled by photographs, especially of face-on galaxies, which are always much more transparent to the eye!

Just look at the drawings of the various galaxies throughout this book. I find it remarkable that I can look at an oblique galaxy such as M98, 55 million light years away, and resolve more detail more easily than I can for a face-on galaxy like M74, which is 23 million light years closer. This is similar to the effect one sees when looking at a planetary nebula's spherical shell of gas, where more light can be seen on the outer edge than in the center because more material is concentrated in our line of sight on the edge.

In 1995 near-infrared images of M100 released by astronomers working on Mauna Kea in Hawaii and on La Palma in the Canary Islands revealed stubby spiral arms close to the nucleus that appear to be rotating in the opposite direction of arms farther out in the giant spiral. The astronomers suggested that these represent regions of new star formation, where hot young stars are being forced into independent orbits. There is a lack of symmetry in the brightest features of M100's disk, which might suggest that the galaxy is being tidally warped by its neighbors. The most surprising image of M100, however, came in late 1993, when the newly repaired Hubble Space Telescope turned its Wide Field and Planetary Camera onto the galaxy's core, revealing a vortex of blue and yellow suns that looks like a galaxy within a galaxy. The HST also resolved

M100

The Mirror of M99
NGC 4321
Type: Spiral Galaxy
Con: Coma Berenices
RA: 12^h 22^m.9
Dec: 15° 49'
Mag: 9.3
SB: 13.0
Dim: 7'.4 × 6'.3
Dist: 55 million l.y.
Disc: Pierre Méchain, 1781

MESSIER: [Observed 13 April 1781] Nebula without a star, with the same luminosity as the preceding one [M99], lying within the ear of corn in Virgo. Seen by M. Méchain on 15 March 1781. These three nebulae, numbers 98, 99, and 100 are very difficult to recognize, because of the dimness of their light. They may be seen only under good conditions, and near meridian passage.

NGC: Very remarkable. Pretty faint, very large, round, very gradually then suddenly brighter toward the middle and to a mottled nucleus. With the 26-inch Leander McCormick refractor, Leavenworth saw M100 as a two-branched spiral.

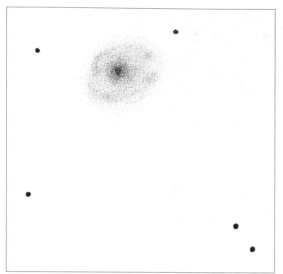

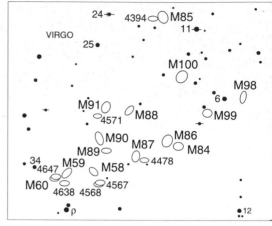

about 20 Cepheid variables in the outskirts of the spiral, leading astronomers to refine the distance scale of the universe. For example, a once generally accepted distance for M99 had been 41 million light years. But the HST data have led to a recalculation of the distance to more like 56 million light years. (The distances of the Virgo galaxies proposed by Tully and used in this book agree with the HST data.)

M101/ M102

(duplicate observation of M101)

NGC 5457
Type: Spiral Galaxy
Con: Ursa Major
RA: 14^h 03^m.2
Dec: +54° 21'
Mag: 7.9
SB: 14.8
Dim: 28'.8 × 26'.9
Dist: 17.5 million l.y.
Disc: Pierre Méchain, 1781; Méchain was also responsible for the duplicate observation (of M102) that same year. He retracted his M102 sighting in 1783.

MESSIER: M101: [Observed 27 March 1781] Nebula without a star, very dark and extremely large, 6 to 7 minutes in diameter, between the left hand of Boötes and the tail of Ursa Major. Difficult to distinguish when the crosshairs are illuminated.

MESSIER: M102: [No date listed] Nebula between the stars o Boötis and ι Draconis. It is very faint. Close to it is a sixth-magnitude star.

NGC: Pretty bright, very large, irregularly round. Gradually, then suddenly, much brighter toward a small, bright central nucleus.

M101 is my second favorite face-on spiral, after M33. Its numerous, far-flung arms and distinct asymmetry make it readily identifiable in photographs. Visually, its pale 9th-magnitude glow is diffuse and difficult to make out, yet it has much to offer if you are patient. It is easy to locate, because you can star-hop to it from the famous visual double star Alcor and Mizar in the handle of the Big Dipper. Southeast of this pair is a path of four 5th-magnitude stars (81, 83, 84, and 86 Ursae Majoris) that leads in the direction of M101, which lies 1½° east-northeast of 86 Ursae at the end of the trail. The galaxy is visible in binoculars, so try them first before taking to the telescope. I wonder whether someone with young eyes could detect it *without* optical aid. At magnitude 7.7 it would be quite a challenge – extremely dark skies and a high-altitude site would probably be required – but the reward is obvious: this person would belong to a club of one, having the distinction of being the only human to see 17.5 million light years into the universe with the naked eye.

Telescopically, at 23 ×, the galaxy's core is compact but slightly elliptical. Knots in phantom spirals emerge from the galactic mists, but mod-

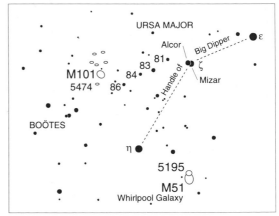

erate power is much more effective in coaxing them out. Low power is best saved for determining the full extent of the far-flung arms. High power, on the other hand, is your best bet for examining the inner region, which shows a sharp nucleus inside a pentagon-shaped core. Only two arms are very definite; beyond that, everything fades in and out of view. You might have to alternate between moderate and low power to preserve your sanity as you try to keep track of their directions. I needed several nights to study this one galaxy, knowing that persistence behind the eyepiece would enhance my view. And, indeed, the time was well spent, because I was able to trace the patterns of a few of its major arms. Still, the view through the eyepiece pales in comparison to the photographic image. And to think, as Timothy Ferris points out in his book *Galaxies,* that billions of suns and interstellar clouds between these arms are too faint to register in photographs. "It is as if we are flying on a landscape at night," he writes, "where the brightly lit cities catch our eye at the expense of the dark farmlands."

For a pleasant diversion, there is a pentagon of 11th-magnitude NGC galaxies 1° north of M101, whose pattern mimics the shape of M101's nuclear region. Another solitary 11th-magnitude galaxy, NGC 5474, lies about 45′ south-southeast of M101; its spectral glow vanishes, however, if the sky is at all hazy.

M103

NGC 581
Type: Open Cluster
Con: Cassiopeia
RA: $1^h 33^m.4$
Dec: $+60° 39'.5$
Mag: 7.4
Dia: 6'
Dist: 8,130 l.y.
Disc: Pierre Méchain, 1781

MESSIER: [no date listed]
Nebula between the stars o
Boötis and ι Draconis. It is
very faint. Close to it is a sixth-
magnitude star.

NGC: Cluster, pretty large,
bright, round, rich, stars of
10th and 11th magnitude.

Stars in an open cluster are destined to be fickle. Without the "glue" of strong mutual gravitation to hold them together – as the hundreds of thousands of stars are held together in a globular cluster – they gradually separate from the main pack and form other alliances. Or, like our sun, travel solo through the emptiness of space. When we look at a loose cluster of stars through a telescope, then, it is difficult to judge what state of association, or disassociation, it is in. The last object in Messier's original catalogue, M103 in Cassiopeia, is a case in point. Astronomer Harlow Shapley went so far as to call this stellar sprinkle a possible chance alignment of stars. But recent data suggest that M103 is a true cluster with as many as 172 stars in an area of only 6′ of arc.

Located about 1° northeast of Delta (δ) Cassiopeia, the cluster is easily spied in 7×35 binoculars, which also resolve the cluster's brighter stars, the brightest being magnitude 10.5. The view through the telescope is really best at low power, when the loose grouping appears more compact. When I place brilliant Delta on the western side of the low-power field of view, M103 lies between it and a congregation of four other open clusters to the east! What a dynamic field, even if these clusters are not related.

M103 is clearly the brightest of the five clusters, and it is easily distinguished by its Christmas-tree shape. The luminous star at the apex of the tree is Struve 131, a double star with 7.3- and 10.5-magnitude components separated by nearly 14″; Smyth saw their colors as straw and dusky blue. Apparently this double is not a true member of the cluster, though.

Smyth also saw a red 8th-magnitude star southwest of the brightest star in the cluster. D'Arrest described this same star shining with a rose

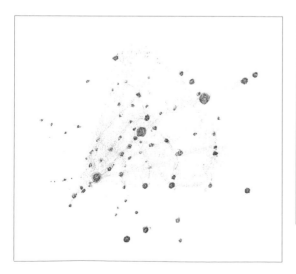

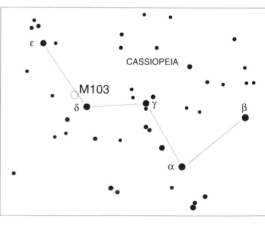

tint, but he recorded it as *10th* magnitude. Luginbuhl and Skiff describe it as "very red" and shining at magnitude 8.5. Photometric measurements confirm that the star is fairly red. And as for me, curiously, I saw no reddish color at all.

M104

Sombrero Galaxy
NGC 4594
Type: Spiral Galaxy
Con: Virgo
RA: 12^h 40^m.0
Dec: −11° 37′
Mag: 8.0
SB: 11.6
Dim: 8′.7 × 3′.5
Dist: 65 million l.y.
Disc: Pierre Méchain, 1781

MESSIER: None

NGC: Remarkable, very bright and large, extremely extended in position angle 92°, very suddenly much brighter toward a central nucleus.

Picturesque M104 is an enigma. In the 1920s Edwin Hubble began classifying spiral galaxies based on the sizes of their central bulge relative to their arms. A galaxy such as M74 in Pisces, for example, has a very tight nuclear region and wide-sweeping spiral arms. M96 is on the other side of the spectrum, with a large bulge but relatively insignificant arms. Like M96, M104 displays a dominant central bulge in long-exposure photographs, but this nearly edge-on system also shows a preponderance of interstellar gas and dust, which is characteristic of a more evolved spiral galaxy, somewhere between M96 and M74 in the classification scheme. Another quizzical finding was made by Vesto Slipher of Lowell Observatory, who in 1913 became the first astronomer to detect rotation in a galaxy other than our own. By studying the spectrum of M104, he deduced that not only was the galaxy receding from us at 700 miles per second, but its disk was actually rotating: one side was moving toward us while the other side was moving away from us.

You can find the Sombrero Galaxy, as M104 is aptly called, hovering about 5½° northeast of Eta (η) Corvi, just across the border in Virgo. My wife picked up M104's bright 8th-magnitude glow while sweeping the region with 7×35 binoculars! Although it is some 20° south of the Virgo Cloud of galaxies, the Sombrero is nonetheless probably an outlying member of that group. Astronomers now believe that the galaxy is receding from us as at more than 600 miles per second and that it is extremely far away – 65 million light years.

Some aficionados would prefer that the Messier catalogue *end* with this stunning object, perhaps feeling that the six subsequent objects (M105–M110) are visually anticlimactic by comparison. So, how does M104 look in a telescope?

Through the 4-inch, low power shows the galaxy as just a tiny oval glow that begs for more magnification. And, indeed, switching to medium

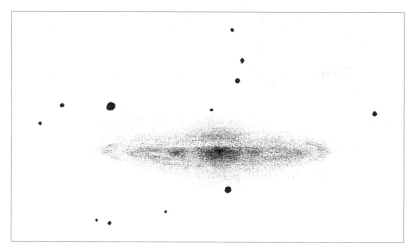

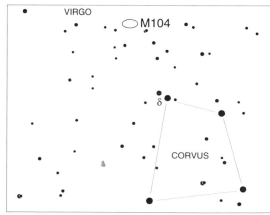

power makes a dramatic difference. The galaxy displays a brilliant core that seems to illuminate the surrounding oval shroud from within, like a distant bonfire seen through a thick fog. A long, bright, needlelike extension runs straight across the major axis of the oval. The sharpness of this line reveals the position of the telltale dark lane, the edge of the Mexican hat's brim. It's quite remarkable to see such detail in a galaxy with only a glance.

With time the brilliant dome over the northern part of the core appears to be straddled by faint condensations, one on either side. The core is the brightest feature of the three, followed by the western knot, and then the eastern one. With averted vision the eastern portion of the Sombrero's brim breaks up and flares into a wide brushstroke of light, which shines more brilliantly than the western portion of the brim. Every now and then the lashlike dust lane wafts into view. High power should reveal the full secrets of these glimmering visions. I find the galaxy looks best at 130×, when it changes from a misty brew of suggestions into a *galaxy* with tightly wound spiral arms and clumps of unresolved starlight. Most surprising is that high power shows the nucleus shining with a yellow light. I have found no other reference to this.

The most challenging details lie in the southern portion of the halo, where a faint dome of light connects to the dark lane. Now switch to low power to see if you can discern a soft light enveloping the entire system. With high power can you resolve the galaxy's individual arms – crescents of light and dark that seem to ripple away from the core? With a little imagination these crescents can help you see the galaxy in three dimensions, like Saturn and its rings when they are nearly parallel to our line of sight. Few objects in the heavens allow users of small telescopes such an interesting visual perspective.

M105

NGC 3379
Type: Elliptical Galaxy
Con: Leo
RA: 10^h 47^m.8
Dec: + 12° 35′
Mag: 9.3
SB: 12.1
Dim: 5′.4 × 4′.8
Dist: 26 million l.y.
Disc: Pierre Méchain, 1781

MESSIER: None

NGC: Very bright, considerably large, round, suddenly brighter in the middle, mottled.

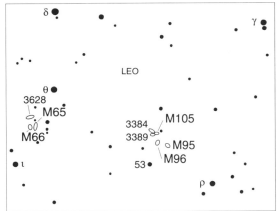

This elegant elliptical, a member of the Leo Group of galaxies, is perhaps the purest object in the Messier catalogue. By that I mean its slightly oval disk shows the least amount of imperfection. Of course, with a diameter of a mere 4′, there's not a lot of room for imperfection. To our eyes M105 lingers only 1° north and slightly east of its traveling companions, M95 and M96. In reality, however, M105 is separated from them by a dizzying 400,000 light years.

Its central core shines brilliantly, "like a 10th-magnitude star," as d'Arrest wrote. Then a bright inner region gives way to an outer halo that fades gradually and uniformly into the cloak of darkness. Only in the very core do I even suspect detail, which amounts to a knot to the southwest of the nucleus and another knot just to the northeast of it; in support of this observation, John Herschel thought he could resolve that glow.

Also traveling with M105 are two other galaxies, NGC 3384 and NGC 3389, both spirals. About 7' northeast of M105, 10th-magnitude NGC 3384 has a round core and dim extensions that quickly taper off to sharp points. NGC 3389 is far more difficult at 12th magnitude, but use high magnification to search for its elusive, crooked arms. All three share the same low-power field with M95 and M96.

M106

NGC 4258
Type: Spiral Galaxy
Con: Canes Venatici
RA: 12^h 19^m.0
Dec: +47° 18'
Mag: 8.3
SB: 13.8
Dim: 18'.6 × 7'.2
Dist: 22 million l.y.
Disc: Pierre Méchain, 1781

MESSIER: None

NGC: Very bright and large, very much extended north–south, suddenly brighter in the middle to a bright nucleus.

In photographs M106 looks like a scarred survivor of galactic violence, and it is. Like M94, this oddly shaped spiral has experienced episodes of violent upheaval in its past. Indeed, we seem to have caught the nucleus in the throes of an enormous explosion, which has strewn several tens of millions of tons of matter across the plane of the galaxy. Its arms are bruised with rich star-forming regions in a mildly chaotic stew of galactic turbulence. Even the galaxy's outer arms look as if they have been stretched like taffy until they nearly snapped free of the main body. Radio images reveal still other arms beyond the limp visual appendages. In 1992 it was announced that, though M106 is seen nearly face on, a disk of gas and dust surrounding the galaxy's core is seen nearly edge on. Then in 1994 a team of radio astronomers discovered a black hole near the galaxy's nucleus. There's more to this Messier object than meets the eye!

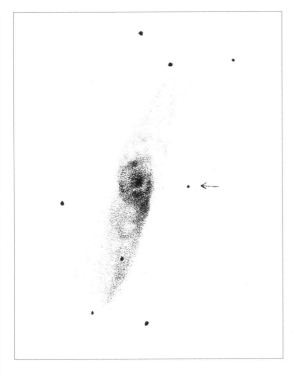

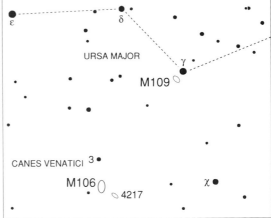

Still, with its belly full of light and two main skeletal arms, M106 is a satisfying view in small-aperture telescopes. In binoculars it glows faintly 2° south of 3 Canum Venaticorum. Its saucer-shaped nuclear region immediately challenges viewers with its complex, mottled texture. The galaxy has essentially the same orientation in space as does the Great Andromeda Galaxy (M31), but only half the mass.

The outer arms of M106 have a distinct S shape and are easy to follow at 23× in the 4-inch. Despite its symmetry, the northern half of the galaxy looks brighter to me than the southern half.

The inner arms also have an S curvature to them, though tighter. So here we have two Ss superposed on one another – one large, one small. Now use high power to study the nuclear region, especially the area around the central condensation. You should see dark matter separating it from the innermost spiral structure. But do you see the tiny S (another one!) of *dark* material centered on the nucleus? It's almost as if all this galactic violence rippled through the entire system in an orderly fashion! Look also for a series of dark patches aligned with the nucleus along the major axis, just north of the bright core.

My drawing of M106 is a composite. It is based on views from three nights of observing with low, medium, and high power. The original sketch I made of it on the third and final night showed a star that I did not record on the other two previous observing sessions. I have indicated the mystery star with an arrow. Curiously, I have not been able to identify this star on photographs of M106. Could this have been a supernova – yet another incidence of violence in this seemingly troubled galaxy?

M107

NGC 6171
Type: Globular Cluster
Con: Ophiuchus
RA: 16^h 32^m.5
Dec: –13° 03'
Mag: 7.8
Dia: 13'
Dist: 19,200 l.y.
Disc: Pierre Méchain, 1782

MESSIER: None

NGC: Globular cluster, large, much compressed, round, well resolved.

M107 is one more example of a globular whose light appears to be affected by interstellar dust. Located 3° south-southwest of 2.6-magnitude Zeta (ζ) Ophiuchi, this globular, from our perspective, sits virtually on top of the hub of the Milky Way over dusty Scorpius. Long-exposure photographs have revealed several possible obscured regions in the cluster. Regardless of the dust, the globular's pale glow can be detected in binoculars, but not with the naked eye (at least I couldn't see it). M107 shines with a magnitude of 7.8 and measures 13' in angular diameter.

Many astronomers theorize that our galaxy began 15 billion years ago as a sphere of gas that eventually collapsed into its present visible form – a central bulge and a disk – and that the ancient globular clusters, which form a spherical halo around the disk, mark the original size of this primordial cloud.

At low power M107 instantly appears elongated in the east–west direction with at least four 11th-magnitude stars around it "in the form of a crucifix," as Smyth recognized. Smyth also inadvertently discovered the dustiness of the surrounding region, calling it a comparative desert. At

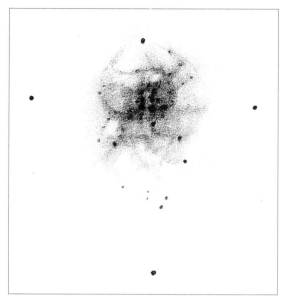

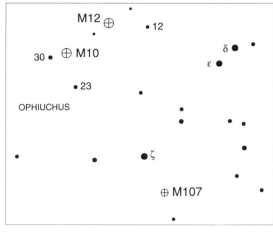

first glance the cluster looks rather loose at $23\times$ and shows no strong central condensation. A prolonged gaze, however, will reveal several fairly bright stars populating the ill-defined outer halo, while the rest of the cluster looks granular.

The big surprise comes with moderate and high powers, when the outlying stars stand out boldly against unresolved hazy wisps, which look like ruffled hair. The nucleus is composed of a boxlike arrangement of stellar patches sliced by a dark lane running from north to south. This dark lane, together with the stars bordering it, remind me, somehow, of a cobblestone path in a morning mist.

M108

NGC 3556
Type: Spiral Galaxy
Con: Ursa Major
RA: 11^h 11^m.5
Dec: +55° 40'
Mag: 10.0
SB: 13.0
Dim: 8'.7 × 2'.2
Dist: 46 million l.y.
Disc: Pierre Méchain,
1781–1782

MESSIER: None

NGC: Quite bright, very large,
very much extended at
position angle 79°, becoming
brighter in the middle,
mottled.

M108 is a tiny spiral galaxy by astronomical standards, only about one-twentieth the mass of M31. Although we see the galaxy nearly edge on, the central bulge is all but absent. Whatever shines in the nuclear region is masked by turbulent eddies of dark matter lining the galaxy's highly fore-shortened arms. The nucleus might have also depleted itself long ago, or it has only periodic bursts of energetic activity that keep it going. Brilliant regions of star formation among the obscuring matter look like signal flares burning in a storm.

On the nights I viewed M108 from the summit of Kilauea volcano, the galaxy was high overhead. So I had to use a star diagonal to relieve the tension in my neck. But that meant I also had to look "down" into the eyepiece. Because I like to observe with both eyes open, I was forced to use a flap on my winter hat to block light from entering the eye I wasn't using. Where was the light coming from? It was the polished glow of the Milky Way reflecting off the ground!

Visible in binoculars as a faint gray streak only two moon diameters (1°) northwest of the Owl Nebula (M97), or 1° southeast of Merak (Beta [β]

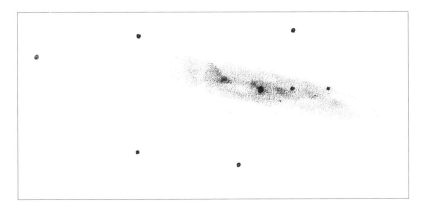

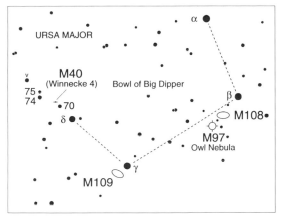

Ursae Majoris), M108 is an interesting spiral. With its highly wrinkled texture, this 10th-magnitude system looks as if it is the shed skin of M82 – a galaxy with a similar demeanor – left here to dry. Even at 23× in the 4-inch, I can recognize M108's elongated (8'.7) and highly mottled disk. But moderate and high powers are needed to show its haphazard array of dark lanes, superposed stars, and irregularly bright nebulous patches, which look like they have been pasted together by a three-year-old.

At 130× one prominent, kinked, dark channel runs along the galaxy's northern rim. More difficult is a long dust lane that follows the southern edge of the galaxy. A beautiful, though delicate, fan of material sprays westward from a faint knot embedded in the western half of M108, and a misty breath of galactic haze shines feebly beneath a 13th-magnitude star in the eastern half. A fainter glow connects that dim patch to the pseudo-nucleus – a fairly bright foreground star near the galaxy's invisible heart. The real challenge is to see the 14th-magnitude bead immediately southwest of that central pip.

M109

NGC 3992
Type: Barred Spiral Galaxy
Con: Ursa Major
RA: 11^h 57^m.6
Dec: +53° 23'
Mag: 9.8
SB: 13.5
Dim: 7'.6×4'.7
Dist: 55 million l.y.
Disc: Pierre Méchain,
1781–1782

MESSIER: None

NGC: Quite bright, very large, pretty much extended, suddenly brighter in the middle to a bright mottled nucleus.

Alas, the Messier catalogue ends not with a bang but a whimper. The last two objects, M109 and M110 are hardly lens-shattering sights. M109 is a typical barred spiral galaxy; in fact, about one-third of all known spirals are of this nature, characterized by a prominent central bar of stars off of which flow more delicate spiral arms. Glowing at 10th magnitude, it can be seen in 7×35 binoculars 1° southeast of the 2.4-magnitude star Phecda (Gamma [γ] Ursae Majoris) in the bottom of the Big Dipper's bowl.

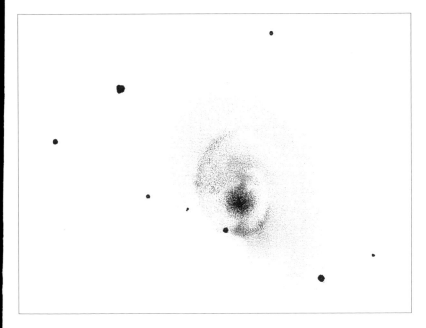

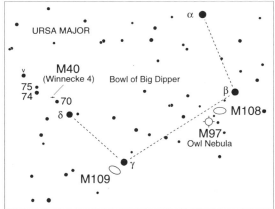

Whenever I sweep toward M109 and its oval shape enters the field, I get a tingle as if I have encountered a comet, as Méchain must have when he first spotted the object. But, because this glow is so close to 2nd-magnitude Phecda, I wonder for a fleeting moment whether I am just seeing a reflection of that star. A good tap on the telescope tube assures me that I am not.

At low power the galaxy displays a well-defined core and a halo muddled with irregularities. Medium power proves it to be a more satisfying barred spiral than either M91 in Coma Berenices or M95 in Leo. The central bar of M109 appears prominent, and the nebulous arcs and hints of spiral arms that emanate from it do not require a strong effort. It is noteworthy, however, that neither Jones nor Clark could detect the bar in an 8-inch telescope; this might say something about the galaxy's appearance in less-than-pristine skies. One night while I was observing it, a brief earthquake caused the scope to jitter, which to my delight brought out some of the galaxy's fainter details – especially the crablike pincers extending to the east. The most difficult task is tracing the pale hazes that make up the outer arms. Also, don't confuse the 12.5-magnitude star 50″ north-north-west of the galaxy's core with a supernova; it is just a foreground star, a resident of our own Milky Way Galaxy.

Swing your scope south-southwest from M109 about 1′ and have a look at NGC 3953, a 10th-magnitude barred spiral similar in appearance to M109.

M110

NGC 205
Type: Elliptical Galaxy
(companion to M31)
Con: Andromeda
RA: 0ʰ 40ᵐ.4
Dec: +41° 41′
Mag: 8.0
SB: 13.9
Dim: 21′.9 × 11′.0
Dist: 2.3 million l.y.
Disc: Messier, 1773

MESSIER: None

NGC: Very bright and large, much elongated in position angle 165°, very gradually brightening to a very much brighter middle.

We now turn our attention to the final addition to the Messier catalogue, an object recommended for inclusion by Kenneth Glyn Jones as recently as 1967. M110 is the larger of the two prominent elliptical satellite galaxies that flank the majestic Andromeda Galaxy (M31), the smaller and closer one being M32. M110 hovers northwest of M31, appearing as a fuzzy oval glow with a total visual magnitude of 8.0 but a much fainter actual surface brightness of 13.9 because of its large size (21′.9 × 11′.0).

High magnification reveals a highly mottled nucleus, with the portion closest to M31 detached by an obvious dust lane! A starlike object lies just south of M110's compact nucleus, and it too seems detached by dark matter, but this might simply be a contrast illusion. Furthermore, the east and west sides of the inner halo exhibit concentrations that are clearly separated from the nucleus. Thus, M110's nucleus appears ringed by arcs of diffuse starlight, which makes the following description of M110 by Heber Curtis of Lick Observatory very intriguing: "The bright central portion . . . [shows] traces of rather irregular spiral structure. Nucleus almost stellar. Two small dark patches near brighter central portion."

Although this peculiar galaxy does not have spiral structure, the dark patches do exist and show on photographs made with large-aperture telescopes. But it is possible for *you* to see them from a dark sky with a 4-inch telescope and high magnification! And M110 tolerates high powers well. See if you get the impression that, like M32, this galaxy has a faint extension or bar running north-south through it.

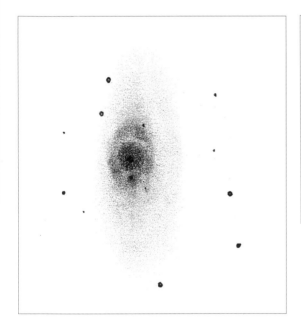

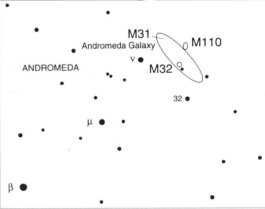

5 Some thoughts on Charles Messier

Charles Messier is best remembered for the objects he largely tried to avoid, and help others to avoid. Messier was first and foremost a comet hunter, and he dedicated clear, moonless nights to this pursuit. He was quite successful in his search, having found some 15 comets, though only 12 bear his name (with some exceptions, comet names were not officially used until the twentieth century). In his time, comet hunting was intensely competitive, and Messier was inspired not only by the search but also by the fame new comet discoveries would bring him. The fuzzy, *non*cometary objects he encountered intrigued him, in part, because they fooled him, if only briefly. Messier created his renowned catalogue largely to benefit *comet hunters*, so they would not confuse these nebulous glows with comets. But it's also possible Messier knew he would reap fame by discovering new nebulae.

Messier's first list of 45 objects, published in the *Memoirs* of the French Academy of Sciences for the year 1771, was well received by his friends and colleagues. In the introduction to the work, astronomer Jerome de Lalande praised Messier, saying the task could only have been undertaken by an indefatigable and experienced observer. William Herschel, in particular, regarded Messier's catalogue as an excellent collection of clusters and nebulae, and it apparently prompted Herschel to undertake his own search for similar deep-sky objects. (Herschel employed much larger telescopes in his survey, which led to the additional discovery of more than 2,000 objects.) Nevertheless, legitimate questions can be raised about why certain objects made it into Messier's catalogue and other seemingly obvious ones did not.

Messier was surprised to resolve stars in several objects described by their discoverers as nebulous patches. But if his aim was to help comet hunters avoid objects that might prove confusing, why would he include *obvious* open clusters, such as M23, M34, M45, and M47? Perhaps, because they appear fuzzy to the naked eye, he did so for the benefit of the naked-eye gazer. (It should be noted that in the mid-1700s spotting a new comet with the unaided eye was not uncommon.)

It would be fruitless to question the observing ability of the man described by King Louis XV as the "ferret of comets." Messier was highly respected for his keen eye and attention to detail. His expertise was not limited to the visual search for comets, as he was an equally skilled observer of, among other things, the planets, stellar occultations, and sunspots. Still, one has to wonder how Messier could have *missed* so many relatively bright nebulous objects during his comet hunts. Anyone who takes the time to sweep the sky with a telescope around the famed M objects cannot help but chance upon some nebulae and clusters that were not included in his catalogue, apparently escaping his gaze. But did they?

We know that of the 109 objects in the catalogue, Messier actually discovered 41 of them; the other objects had been found by others, but Messier recorded their positions and appearance anyway since they could be confused with comets. Once he started the catalogue, Messier wrote, he endeavored to find other nebulosities; he made this something of a project for its own sake, at least in the beginning. But after the first catalogue of 45 objects was published, most of the objects recorded thereafter by Messier were found serendipitously, when particular comets he was following would happen to pass near a nebulous patch. On average, he discovered only about two new catalogue objects per year.

One particularly puzzling omission from the catalogue is the Double Cluster in Perseus. Why would Messier not include this conspicuous Milky Way object, which appears distinctly cometlike to the naked eye? Yes, the cluster does resolve with the slightest optical aid into a myriad of individual stars, but the same could be said of the Pleiades, the brilliant and easily resolvable open cluster in Taurus, which he catalogued as his 45th entry. His inclusion of the Pleiades makes it hard to argue that he considered some objects too obviously *not* cometlike in appearance to be confused by them, or to bother recording them.

One can only speculate about the absence of the Double Cluster and other bright objects of which we presume Messier should have taken notice, even with his rudimentary scopes. And one can perhaps detect a bit of defensiveness in Messier about his catalogue in some remarks he made in the French almanac *Connaissance des Temps* for 1801, just a few years before his death. "After me, the celebrated Herschel published a catalogue of 2,000 which he has observed. This unveiling of the sky, made with instruments of great aperture, does not help in a perusal of the sky for faint comets. Thus my object[ive] is different from his, as I only need nebulae visible in a telescope of two feet [length]. Since the publication of my catalogue I have observed still others; I will publish them in the future . . . for the purpose of making them more easy to recognize, and for those searching for comets to remain in less uncertainty." It would be informative to know Messier's feelings on this 30 years earlier. In *Messier's Nebulae and Star Clusters* Jones comments that the argument that large apertures do not help in the discovery of comets doesn't hold water. But I find Messier's remark about his objective being different from Herschel's, and his need to see only nebulae visible in a telescope of two-foot length, as testimony to what I believe was his unwavering purpose, the pursuit of comets.

It is also surprising that Messier could not telescopically resolve the great globular clusters M5 in Serpens Caput and M13 in Hercules. Of course, he was limited not only by the poor location of his observatory but also by the poor quality of the optics in his instruments. He used about a dozen different telescopes ranging from reflectors with inefficient speculum mirrors (the largest having an 8-inch aperture) to simple refractors with apertures of up to 3½ inches. He generally employed magnifications of between 60× and 138×. Compared to today's instruments, all these telescopes were of substandard optical quality. "It is a pity," wrote Smyth in his *Cycle of Celestial Objects*, "that this active and assiduous astronomer could not have been furnished with one of the giant telescopes of the present day.... One is only surprised that with his methods and means, so much was accomplished." Likewise, Lalande could only lament that the precise and zealous Messier could not live under purer and less cloudy skies, which obviously hampered his ability to find new comets and nebulae.

Some of today's skilled comet hunters claim to have an innate sense of what a new comet will look like through the telescope. Thus, it is the comet hunter's prerogative to skip over or ignore whatever object he or she deems not cometary. It is reasonable to presume that Messier had his own criteria for judging whether an object was sufficiently cometlike in appearance to warrant noting its position, though he did not relate those criteria. But it should be said that he never endeavored to conduct a comprehensive visual search for noncometary nebulae. An old Chinese proverb tells us that obstacles are what we see when we take our mind off the goal. And Messier's goal was to discover comets, not nebulae or star clusters. In contrast, Herschel's goal *was* to systematically survey the entire sky for deep-sky bounty beyond the limits of Messier's instruments. Most of the relatively bright non-Messier objects we enjoy today (at least from the Northern Hemisphere) were discovered by Herschel. There can be no doubt that had Messier conducted a similar deliberate search for noncometary objects, his catalogue would have turned out vastly larger and less haphazard than it is. But that was not Messier's concern, comets were. It is only by the fickleness of fate that none of his comets had the staying power of his catalogue of deep sky wonders.

6 Twenty spectacular non-Messier objects

Without question, Messier's catalogue contains many of the sky's brightest and most glorious deep-sky objects. But there are a host of other exquisite objects that he did not include, either because they were too far south to be seen from his observing location, or because he missed, ignored, or somehow overlooked them. The following are brief descriptions of 20 of my favorite non-Messier objects, in no particular order. I only hope these objects will inspire you to continue scanning the skies for your own celestial treasures.

1

Why Messier failed to include the Perseus Double Cluster in his catalogue is a mystery. This stunning object is among the finest celestial showpieces in the northern sky. Riding high in the winter sky midway between Delta (δ) Cassiopeiae and Gamma (γ) Persei, this pair of stellar islands looks to the naked eye like a 4th-magnitude nebula spanning 1½ moon diameters. Binoculars separate the two clusters, NGC 869 and NGC 884, which are ½° apart, but a rich-field telescope provides the best view, revealing some 200 suns piled like rubies and diamonds on black velvet. Strings of stars stretch between the couplet, like arms entwining them in an eternal embrace. With an estimated age of only a few million years, the Double Cluster is one of the youngest galactic clusters known.

NGC 869/884
Double Cluster
Type: Open Cluster
Con: Perseus
RA: 02^h 19^m.0 (NGC 869), 02^h 22^m.5 (NGC 884)
Dec: + 57° 08′ (NGC 869), + 57° 07′ (NGC 884)
Mag: 3.5 (NGC 869), 3.6 (NGC 884)
Dia: 1.5° (each)
Dist: 7,100 l.y. (NGC 869); 7,500 l.y. (NGC 884)

2

A million sparkling suns packed into an oval disk slightly larger than the full moon's, that's globular cluster Omega Centauri, arguably the finest specimen of its kind. Shining at magnitude 3.5, Omega Centauri appears to the naked eye as a "fuzzy star" or the head of a comet. It is best seen from the southern half of the United States and points farther south; Messier could not have discovered it from Paris; otherwise, he undoubtedly would have praised its beauty. The globular looks like a sparkling ball of light in binoculars, while a 4-inch telescope shows myriad blue suns bursting out of an ever-tightening core seems to scintillate with nervous energy. Look for two dark patches on the core.

NGC 5139
Omega Centauri
Type: Globular Cluster
Con: Centaurus
RA: 13^h 26^m.8
Dec: –47° 29′
Mag: 3.9
Dia: 53′
Dist: 18,250 l.y.

3

NGC 3372
Eta Carinae Nebula
Type: Diffuse (Emission)
Nebula
Con: Carina
RA: 10^h 43^m.8
Dec: –59° 52′
Dia: 3°
Mag: 3.0
Dist: 10,000 l.y.

Unrivaled in nebulous splendor, the Eta Carinae Nebula is a southern treasure that surpasses even the great Orion Nebula in size and beauty. It is a celestial continent of bright and dark nebulosity spanning more than 2° of sky. To the naked eye it looks like one of many hazy-looking objects populating the fabulously rich section of the southern Milky Way between Vela and the Southern Cross. But its inconspicuous naked-eye appearance belies the grandeur that awaits telescopic observers: brilliant clouds of twisted gas mix playfully with dark clouds of obscuring matter, one part of which mimics the appearance of a black keyhole. The vaporous swirls in this complex nebula surround one of the sky's most mysterious stars – Eta Carinae, a novalike object 150 times larger and 4 million times brighter than our sun. Although Eta Carinae now shines around 6th magnitude, the star is known to experience violent eruptions that cause it to brighten to magnitude –0.8, outshining every star in the sky except for Sirius. That happened last in 1843.

4

NGC 6231
Type: Open cluster
Con: Scorpius
RA: 16^h 54^m.0
Dec: –41° 47′
Mag: 2.6
Dia: 14′
Dist: 5,800 l.y.

Somehow, this brilliant 2.6-magnitude open cluster in the tail of Scorpius went unnoticed by Messier. I find that surprising, because it is the most cometlike naked-eye spectacle in the heavens that is not a comet – especially when you include the elongated cluster known as Harvard 12 or Trumpler 24 embracing it to the north. Together, NGC 6231 and Harvard 12 hang above the southern horizon like a bright comet with a broad dust tail. Although these objects have separate catalogue identities, their stars meld; that's because they belong to the same group of high-luminosity stars known as the "Sco OB1 Association," which marks the location of one of our galaxy's spiral arms – the one closer to the galactic center than the one containing our sun. Certainly, these objects would have been appropriate for Messier's catalogue had he spotted them.

5

NGC 2477
Type: Open Cluster
Con: Puppis
RA: 07^h 52^m.3
Dec: –38° 32′
Mag: 5.8
Dia: 20′
Dist: 4,200 l.y.

One of my all-time favorite binocular objects, and yet another surprising omission from Messier's catalogue, is this 6th-magnitude open cluster. Located about 2½° northwest of 3.3-magnitude Zeta (ζ) Puppis, NGC 2477 is the brightest open cluster in that cluster-laden constellation and can be seen with the unaided eye on clear, dark nights. Through binoculars its compact form appears strikingly like that of a tailless comet, one about the size of the full moon! Telescopically, the cluster is a tight, almost globularlike swarm of some 300 glinting gems. In his book *Star Clusters* the late Harlow Shapley said NGC 2477 was either a massive open cluster that is 500 million to 1 billion years old or the loosest of globular clusters.

For those who can dip deep into southern skies, NGC 6397 is a most pleasing target for small telescopes. Under dark skies it is visible to the naked eye as faint "star" on the eastern fringe of the Milky Way, about 10½° south of Theta (θ) Scorpii, a 1.9-magnitude Type-F in the Scorpion's tail. Whenever, for pleasure, I'm just scanning that region, which is rich in clusters, the alluring glow of this 6th-magnitude wonder inevitably captures my attention. Through a 4-inch refractor the globular is a clean wash of loosely packed starlight. Many of its members are easily resolved – about two dozen of them shine between magnitude 10 and 12. The cluster also contains many bright red-giant stars that are 500 times more luminous than our sun. At a distance of 7,100 light years, NGC 6397 may be the closest globular cluster to our solar system – some 2.5 times closer than the great Omega Centauri cluster.

NGC 6397
Type: Globular Cluster
Con: Ara
RA: 17^h 40^m.7
Dec: –53° 40′
Mag: 5.3
Dia: 30′
Dist: 7,100 l.y.

An enormous galaxy with three times the number of stars that reside in our Milky Way Galaxy, this celestial giant can be located with binoculars 4½° north of Omega Centauri. You'll find it shining as a conspicuous 7th-magnitude glow with an apparent diameter equal to that of the full moon. Through a telescope NGC 5128 looks like two halves of a broken egg – its whitish shell cracked open by a black absorption nebula. When I first swept up this magnificent object in the 4-inch, I thought I had encountered a comet with a strong parabolic hood: the dark lane looked like the shadow of that "comet's" nucleus running down the length of its broad tail. Had Messier found this object, I am certain he would have included it in his catalogue. The galaxy itself is a strong radio emitter, producing a signal 1,000 times more intense than that of our own Milky Way. Indeed NGC 5128 is among the most peculiar galaxies known, with an explosive nucleus that has jettisoned millions of solar masses of material into space. Astronomers believe Centaurus A is the remains of a cosmic collision between two galaxies, an elliptical and a spiral.

NGC 5128
Centaurus A (Radio Source)
Con: Centaurus
Type: Peculiar Galaxy
RA: 13^h 25^m.5
Dec: –43° 01′
Mag: 6.7
Dim: 31′ × 23′
Dist: 15 million l.y.

8

NGC 3242
Ghost of Jupiter
Type: Planetary Nebula
Con: Hydra
RA: 10^h 24^m.8
Dec: –18° 38'
Mag: 7.8
Dia: 16"
Dist: 3,300 l.y.

Here is another unfortunate "miss" by Messier and his contemporaries, because undoubtedly NGC 3242 in Hydra is one of the finest examples of a planetary nebula in the heavens. It shines a full magnitude brighter than M57 (the famous Ring Nebula in Lyra), has an 11.4-magnitude central star within range of the smallest of telescopes, and sports a pale blue disk that is about the same apparent size of Jupiter (thus the Ghost of Jupiter nickname). You can locate NGC 3242 with binoculars nearly 2° south of Mu (μ) Hydrae. Moderate-size telescopes will reveal its ring structure embedded in a larger oval disk of faint light. From dark skies, all its principal features can be seen with a 4-inch telescope. Unlike some planetaries, this one has a high surface brightness, so the nebula can withstand exploration with high magnification. The true diameter of the outer shell may be about 0.6 light year.

9

NGC 253
Type: Spiral Galaxy
Con: Sculptor
RA: 00^h 47^m.6
Dec: –25° 17'
Mag: 7.6
Dim: 30'.0 × 6'.9
Dist: 10.5 million l.y.

For small-telescope users the highly elongated spiral galaxy NGC 253 in Sculptor is rivaled in detail only by M31, the Great Andromeda Galaxy. Although NGC 253 shines only as bright as a 7.6-magnitude star (whose light is spread over an area of 30') its high surface brightness makes it a grand sight through any aperture and with most magnifications. The galaxy displays a highly uniform, though mottled texture. Splashes of dark dust are interspersed with the bright patches. Especially prominent is one dark lane to the west of the nucleus. NGC 253 is the brightest member of the Sculptor Group of galaxies – the closest galaxy cluster to the Local Group, which includes the Milky Way – and is roughly the same size as M31. The object lacks the obvious, starlike core typically seen in many other spiral systems. Indeed, astronomers have found that this galaxy's nucleus seems to be simmering, with gases flowing rapidly away from the nuclear region.

10

NGC 6960 and NGC 6992
Veil Nebula or *Cirrus Nebula*
Type: Supernova Remnant
Con: Cygnus
RA: 20^h 45^m.7 (NGC 6960)
Dec: + 30° 43' (NGC 6960)
Mag: 7th, but low surface brightness
Dia: 2.5°

Commonly referred to as the Veil Nebula, the dual arcs of NGC 6960 and NGC 6992 are among the most sought-after sights at summer star parties. In long-exposure photographs, several extended, arcing streamers appear to float against the Milky Way surrounding the star 52 Cygni. NGC 6960, the brightest arc, can be seen easily in binoculars from a dark sky, and both of these NGC nebulae are within range of 7 × 35 binoculars. They appear ghostly in form, like pale images of fractured chicken bones. The Veil streamers are the detritus of a star that exploded some 40,000 years ago.

With more than 50 bright and colorful suns packed into an area of sky only 10′ across, NGC 4755 ranks high as one of the most spectacular celestial treasures in the southern sky. The cluster is visible to the naked eye as a slightly swollen 4th-magnitude star (Kappa [κ] Crucis) in the most sought-after southern constellation, Crux, the Southern Cross. Binoculars resolve at least nine of its members, which range in brightness from 5.7 to 10.5 magnitude. Small telescopes offer viewers a rich assortment of stellar jewels that shimmer with opalescent light, like a cluster of pearls of various sizes. The bright central region of the Jewel Box Cluster measures about 25 light years in diameter, while a region spanning twice that distance is populated with fainter stellar members. Its nickname comes from an observation by John Herschel, who called it a "superb piece of fancy jewellery."

11

NGC 4755
Kappa Crucis (Jewel Box)
Type: Open Cluster
Con: Crux
RA: 12^h 53^m.6
Dec: −60° 21′.4
Mag: 4.2
Dia: 10′
Dist: 7,600 l.y.

Massive and black, the Coalsack Nebula blots out 26 square degrees of Milky Way just east of Alpha (α) Crucis in the Southern Cross; it also lies just south of the rich Jewel Box Cluster. The juxtaposition of these two disparate objects seems almost ironic: to the naked eye the Coalsack looks like the silhouette of a black hole that is greedily consuming the space around it. By comparison the Jewel Box Cluster looks like a cache of stars that have been plucked from the Milky Way. The black cloud lies some 500 or 600 light years away and is one of the closest dark nebulae to our solar system. Telescopically, the cloud looks shredded, as if millions of dark vapors are wending their way though a forest of dim suns. Seen together with the Jewel Box and the Southern Cross, the Coalsack creates one of the most awe-inspiring sights in the heavens.

12

Coalsack Nebula
Type: Dark Nebula
Con: Crux
RA: 12^h 53^m
Dec: −63° 30′
Mag: –
Dim: 7° × 5°
Dist: 550 l.y.

Nearly 2° east of 17 Comac Berenices lies NGC 4565, the largest and most famous edge-on spiral galaxy in the night sky. In the 4-inch it is not a stunning sight but an elegant one. Shining at 10th-magnitude, the galaxy appears as a slim streak of light with a hazy central bulge that is punctuated by a starlike core. The challenge in small telescopes is to see and trace the dark dust lane that runs along the entire length of this spindle. In photographs taken with large-aperture instruments, this 90,000-light-year-long spiral, which is tilted a mere 4° from edge on, displays dark arcing festoons of dust silhouetted against the bright central bulge. This material was ejected hundreds of light years out of the plane of the galaxy, but gravity is drawing it back in.

13

NGC 4565
Type: Spiral Galaxy
Con: Coma Berenices
RA: 12^h 36^m.3
Dec: + 25° 59′
Mag: 9.6
Dim: 14′.0 × 1′.8
Dist: 20 million l.y.

14

NGC 891
Type: Spiral Galaxy
Con: Andromeda
RA: 02^h 22^m.6
Dec: + 42° 21′
Mag: 9.9
Dim: 13′.0 × 2′.8
Dist: 43 million l.y.

Seen exactly edge on, NGC 891 is the unequivocal rival of NGC 4565 in Coma Berenices (just described). However, the lower surface brightness of this 9.9-magnitude galaxy, which lies 3½° east of the fine double star Gamma[1,2] ($\gamma^{1,2}$) Andromedae, makes it a less attractive sight for small-telescope users. But I suppose my loyalty to NGC 891 stems from childhood, because an image of the galaxy used to appear in the closing moments of the television classic *Outer Limits,* which had a somewhat mystical effect on me. Like NGC 4565, NGC 891 is disrupted by a thick lane of dark matter. In 1940, Carl Seyfert (of *Seyfert galaxy* fame) discovered that the galaxy's dark lane only *appears* dark because its light is up to a magnitude fainter than the surrounding brightness of the galaxy.

15

NGC 2024
Lips Nebula or *Flame Nebula*
Type: Diffuse (Emission)
Nebula
Con: Orion
RA: 05^h 41^m.9
Dec: –01° 51′
Mag: –
Dim: 30′ × 30′
Dist: –

This wonderful but overlooked diffuse nebula in Orion's belt is often portrayed as a challenging telescopic object, but it really is not. True, its pale light is spread over an area as large as the full moon, and it lies a mere 15′ northeast of the blazing 1.8-magnitude Zeta (ζ) Orionis. But from a good dark-sky site it shows clearly in 7 × 35 binoculars. Through the 4-inch, this softly glowing cloud is parted by a thick, serrated dark channel. Thus, NGC 2024 looks very much like a slightly open pair of lips. Despite its distinctive telescopic appearance, I prefer the binocular view because the nebula looks devilishly subtle, like the mysterious smile of the woman portrayed in Leonardo da Vinci's Mona Lisa.

16

NGC 2451
Type: Open Cluster
Con: Puppis
RA: 07^h 45^m.4
Dec: –37° 57′
Mag: 2.8
Dia: 50′
Dist: 850 l.y.

Of the seemingly endless open clusters distributed throughout the Puppis region of the Milky Way band, NGC 2451 has found a special place in my heart. It is a brilliant 3rd-magnitude object 1½ moon diameters across with at least 30 colorful stars within the range of binoculars. These stars seem to burn in adoration of the brightest cluster member – the blazing red 3.6-magnitude c Puppis. With a little imagination, the brightest stars can be seen forming a pattern reminiscent of a scorpion with outstretched claws and raised tail. The cluster almost hugs the horizon from midnorthern latitudes, but it is a stunning sight under southern skies.

A fine 9th-magnitude planetary for small telescopes, NGC 2392 looms in the night sky like a pale apparition on the eastern outskirts of the Gemini Milky Way. It is easily located about 2½° southeast of Delta (δ) Geminorum and 1'.6 south of an 8th-magnitude field star. With low power it is an abstruse sight, appearing to be little more than a star surrounded by dim haze. But increased magnification and a little patience bring out some titillating details. Most noticeable is the planetary's 10.5-magnitude central star, which seems to burn through a dense 15"-wide shell of greenish blue gas. With patience, that diffuse shell might appear mottled with dark patches. Photographs show these best; they also reveal another ring of gas some 40" away. When seen together, the dark patches in the inner shell and the wispy ring of the outer shell suggest the face of an Eskimo, whose dark eyes are peering out of the hood of a fur-lined parka. As with most planetaries, with averted vision the gas surrounding the bright central star appears to swell. The shell, by the way, is expanding at a rate of about 68 miles per second, so its apparent size is growing about 1" every 30 years. It probably first left the Eskimo's central star some 1,700 years ago, making this planetary one of the youngest known.

17

NGC 2392
Eskimo Nebula
Type: Planetary Nebula
Con: Gemini
RA: 07^h 29^h.2
Dec: + 20° 55'
Mag: 9.2
Dia: 40"
Dist: 3,000 l.y.

Although Collinder 399 is visible to the naked eye as a 4th-magnitude fuzzy patch twice the diameter of the full moon, it looks best in binoculars. Finding it entails hopping just 4° northwest from Alpha (α) Sagitta in the summer Milky Way. The shape of this possible open cluster's dozen or so brightest members looks irresistibly like a coathanger; it is one of the few stellar groupings in the sky that immediately looks like its nickname, the Coathanger. All told, this 200-million-year-old aggregation contains about 40 stars without a hint of concentration. Its brightest member shines at 5th magnitude and is visible to the naked eye.

18

Collinder 399
Brocchi's Cluster or
Coathanger Cluster
Type: Open Cluster or
Asterism
Con: Vulpecula
RA: 19^h 25^m.4
Dec: + 20° 11'
Mag: 3.6
Dia: 90'
Dist: 423 l.y.

19

NGC 7000
North America Nebula
Type: Diffuse (Emission)
Nebula
Con: Cygnus
RA: 20^h 58^h.8
Dec: + 44° 20′
Mag: –
Dim: 2°.0 × 1°.7
Dist: 1,600 l.y.

Just 3° east of blue, 0.1-magnitude Deneb (Alpha [α] Cygni) lies NGC 7000 – the unsung hero of the Cygnus Milky Way. In photographs it appears as a sprawling cloud of glowing gas stars bordered by inky opacity, covering an area four times larger than the full moon, and resembling in shape the outline of the North American continent. One misconception, however, is that the nebula itself is visible only in photographs; this is not the case. In fact, it was discovered visually. Its presence can be inferred with the naked eye as an area of enhanced brightness. Binoculars show it unmistakably as a distinct fan-shaped glow with a faint tail to the south, bordered by a dark gulf to the west. A wide-field telescope under a dark sky brings out the full glory of the nebula, including the dark nebulosity forming the "Gulf of Mexico," and the "East" and "West Coasts." A nebula filter will boost the contrast of this low-surface-brightness emission nebula, making it easier to discern.

20

The Milky Way

How ironic that I have spent so much time looking for faint details in distant galaxies with my telescope when the most majestic of all galaxies – the Milky Way – reveals itself to me each night with a beauty, shape, and richness of texture that no telescope or spacecraft can capture as well as our naked eyes can.

Objects Messier could not find

At the end of his catalogue in the *Connaissance des Temps* for 1784, Messier included a list of objects reported by other astronomers that he had been unsuccessful in locating himself. The list, translated from French by Storm Dunlop, follows. As was his custom, Messier refers to himself in the third person.

Nebulae discovered by various astronomers, which M. Messier has searched for in vain.

Hevelius, in his *Prodronie Astronomie*, gives the position of a nebula located at the very top of the head of Hercules at right ascension 252° 24′ 3″, and northern declination 13° 18′ 37″.

On 20 June 1764, under good skies, M. Messier searched for this nebula, but was unable to find it.

In the same work, Hevelius gives the positions of four nebulae, one in the forehead of Capricornus, the second preceding the eye, the third following the second, and the fourth above the latter and reaching the eye of Capricornus. M. de Maupertuis gave the position of these four nebulae in his work *Figure of the Stars*, second edition, page 109. M. Derham also mentions them in his paper published in *Philosophical Transactions*, no. 428, page 70. These nebulae are also found on several planispheres and celestial globes.

M. Messier searched for these four nebulae: namely on 27 July and 3 August, and 17 and 18 October 1764, without being able to find them, and he doubts that they exist.

In the same work, Hevelius gives the position of two other nebulae, one this side of the star that is above the tail of Cygnus, and the other beyond the same star.

On 24 and 28 October 1764, M. Messier carefully searched for these two nebulae, without being able to find them. M. Messier did indeed observe, at the tip of the tail of Cygnus, near the star π, a cluster of faint stars, but its position was different from the one reported by Hevelius in his work.

Hevelius also reports, in the same work, the position of a nebula in the ear of Pegasus.

M. Messier looked for it under good conditions during the night of 24 to 25 October 1764, without being able to discover it, unless it is the nebula that M. Messier observed between the head of Pegasus and that of Equuleus. See number 15 in the current catalogue.

M. l'abbé de la Caille, in a paper on the nebulae in the southern sky, published in the Academy volume of 1755, page 194, gives the position of a nebula that resembles, he says, the nucleus of a small comet. Its right ascension on 1 January 1752 is 18° 13' 41" and its declination is –33° 37' 5".

On 27 July 1764, under an absolutely clear sky, M. Messier searched for this nebula without success. It may be that the instrument that M. Messier was using was not adequate to show it. Subsequently seen by M. Messier. See number 69.

M. de Cassini describes in his *Elements of Astronomy,* page 79, how his father discovered a nebula in the space between Canis Major and Canis Minor, which was one of the finest that could be seen through a telescope.

M. Messier has looked for this nebula on several occasions, under clear skies, without being able to find it, and he assumes that it may have been a comet that had either just appeared, or was in the process of fading. Nothing is so like a nebula as a comet that is just beginning to be visible to instruments.

In Flamsteed's great catalogue of stars, the position of a nebula is given as being in the right leg of Andromeda, at right ascension 23° 44' and polar distance [declination] +50° 49' 15".

M. Messier searched for this on 21 October 1780, using his achromatic telescope, without being able to find it.

Messier marathons

Each spring amateur astronomers around the world run a marathon – not the grueling 26-mile-long race, but a visual race through the night sky to glimpse all 109 Messier objects in a single night. Messier marathons, as they are called, are held during late March and early April, the only time of the year when all the Messier objects are visible between dusk and dawn, and the only time of year when the sun lies in a region of sky devoid of these celestial treasures. On the first clear, moonless night, marathoners start searching low in the western sky at dusk, then hop from one M object to the next, until they've exhausted all the objects (or themselves) by dawn. Aside from clear skies, success requires a decent knowledge of the stars and constellations, efficient use of a telescope, and the ability to read star charts and confirm the appearance and position of each Messier object. These skills prove critical especially during twilight hours, when inevitably some targets must be found.

To the best of my knowledge the Messier marathon originated in Spain when a group of amateur astronomers set out to attempt the task in the 1960s. Several amateur astronomers across the globe independently conceived the notion a decade later. In the March 1979 issue of *Sky & Telescope* magazine, the late Deep-Sky Wonders columnist Walter Scott Houston described how, in the United States, Tom Hoffelder of Florida began marathoning with amateurs in the mid-1970s, while the Amateur Astronomers of Pittsburgh independently started the activity in 1977. On the West Coast, Donald Machholz, a leading comet discoverer, dreamed up the marathon idea in September 1978. Needless to say, serendipity struck like wildfire, and the activity spread worldwide. Today astronomy clubs routinely sponsor Messier marathons, as well as variations on the theme, such as binocular marathons, CCD-imaging competitions, and CCD vs. visual showdowns. One club even puts on a Messier event in which amateur astronomers compete in a cycling race, stopping occasionally to hunt down M objects with portable telescopes!

Messier marathons are popular for several reasons. Besides being fun and challenging, they provide astronomy clubs with an annual activity for their members, and they provide a "proving ground" for testing your observing techniques and search methods. And, let's face it, successful completion of a Messier marathon, just like the running race, earns you "bragging rights" – and possibly a T-shirt or lapel pin. But also, participants often come away from the mad dash around the heavens having learned something useful, whether it be about the sky, their telescope, or themselves.

Which brings me to why I have never participated in a Messier marathon. It's not that there is anything wrong with the activity – as I have mentioned, it sounds like a lot of fun, especially if done with friends. But, as you have seen in this book, I much prefer to dote on objects, often spending hours at a time studying a single one – which would put me at a definite disadvantage in a race! To me it is a philosophical thing, and maybe I'm just a hopeless romantic. But dashing around the sky simply to tally how many of these extraordinary cosmic marvels you can spot in a night is tantamount to sprinting through the halls and galleries of the Louvre just to say you've seen all its famous masterpieces.

But if you are so inclined, and are interested in learning more about Messier marathons, I encourage you to contact your local planetarium or astronomy club, many of which sponsor such events. You might also get a copy of the book by Don Machholz titled *Messier Marathon Observer's Guide: Handbook and Atlas,* which offers suggestions on how to prepare for and conduct a marathon, and contains other useful information.

If you would prefer to take your time logging observations of the Messier objects, but would like to receive recognition for your efforts, organizations such as the Astronomical League (Contact: Berton Stevens, 2112 Kingfisher Lane E., Rolling Meadows, IL 60008) and the Royal Astronomical Society of Canada (Contact: National Office, 136 Dupont St., Toronto, ON M5R 1V2, Canada) have Messier "clubs" you can join. Members are awarded certificates upon completion of the Messier list, and there are lists and awards for both binocular and telescope users.

In addition to hosting Messier marathons, astronomy clubs should also consider using March and the Messier objects to launch larger programs designed to educate the public about astronomy. By celebrating a Messier Month, clubs could bring the magic of Messier and his catalogue to schoolchildren and the greater public through slide shows and special observing events. The idea, of course, would be to showcase the Messier objects with the simple message that the splendors of the universe are readily accessible to anyone who enjoys gazing into the night sky – whether it's with the naked eye, binoculars, or a telescope. Messier-specific activities could run in conjunction with Astronomy Day celebrations, which are held around the same time of year. And there are ample products to distribute or display to inspire the public about these deep-sky splendors, one being the fine Messier Objects poster produced by Sky Publishing Corp., which displays photographs of all the M objects and lists their locations, magnitudes, and other data. As part of any Messier celebration, special training sessions could be held, in which skilled observers not participating in a marathon would help beginners to locate M objects.

A quick guide to navigating the Coma–Virgo cluster

Probing the depths of the Coma–Virgo Cluster may seem like a daunting task. Just look at all of those tiny galaxy symbols jammed together near the center of the wide-field map at the back of this book. But don't let the clutter discourage you. The cluster is actually quite easy to navigate. The simplest approach is to start with M58, one of the brightest members of the Virgo Cluster.

First locate the 3rd-magnitude star Epsilon (ε) Virginis. Five degrees to its west is 5th-magnitude Rho (ρ) Virginis; Rho is easy to confirm in binoculars, because it is the brightest star in the middle of an arc of three stars oriented north–south. Two degrees west of Rho you will find the star 20 Virginis. M58 forms the northern apex of an equilateral triangle with Rho and 20 Virginis. M58 is easy to confirm because it is only 7′ to the west of an 8th-magnitude star.

M58 is a member of what I call the "Great Wall of Galaxies" – a strong line of six Messier galaxies, oriented slightly northwest to southeast, that spans 5° of sky. Pairs of galaxies punctuate either end of the Wall, making identification of these objects easy. Using M58 as your reference point, move about 1½° to the east and slightly south, where you will find your first pairing, M59 and M60, separated by only 30′. M60, which is the brightest galaxy in this string, has a very faint companion, NGC 4647, to the north. Now return to M58 and continue that line an equal distance to the north-west; there you will come upon the bright, round galaxy M87. One degree farther is the other galaxy pairing, M84 and M86.

The Great Wall also forms the baseline of a coathanger asterism of galaxies, which includes M88, M89, M90, and M91. On the chart, notice that M90 forms the northern apex of an equilateral triangle with our reference galaxy, M58, and M87. Furthermore, M89 is near the center of that triangle. However, because M90 is the more obvious of the two, try for it first. Not only is it bright, but it is an oblique spiral galaxy and clearly looks different from the other, more elliptical, hazes.

To find M88, simply move the telescope 1½° to the northeast of M90. M88 is easy to identify, because it is another fine spiral and there is an obvious double star at its southeastern tip.

M91 lies less than 1° due east of M88. But be careful here not to mistake M91 for NGC 4571 just to its southeast. Because there are no other galaxies in this immediate region, you can identify M91 by moving the telescope to the southeast to pick up NGC 4571 (or vice versa).

The final three galaxies – M98, M99, and M100 – should present no problems, because they reside near three binocular stars, the brightest of which is 6 Coma Berenices. All you have to do is locate those stars with your binoculars, point your telescope to them, and you're home free. See how simple it can be?

Suggested reading

I highly recommend the following books and magazines, some of which I referred to throughout this book. They offer additional insights into the Messier objects, observing with a telescope, and the hobby of amateur astronomy in general.

Beyer, Steven L. *The Star Guide: A Unique System for Identifying the Brightest Stars in the Night Sky.* Boston: Little, Brown & Co., 1986.

Burnham, Robert, Jr *Burnham's Celestial Handbook.* Volumes 1–3, 2nd ed. 3 vols. Mineola, N.Y.: Dover Publications, 1978.

Clark, Roger. *Visual Astronomy of the Deep Sky.* Cambridge, Mass.: Sky Publishing Corp.; Cambridge: Cambridge University Press, 1990.

Cragin, Murray, James Lucyk, and Barry Rappaport. *The Deep Sky Field Guide to Uranometria 2000.0.* Richmond: Willmann-Bell, 1993.

Dickinson, Terence, and Alan Dyer. *The Backyard Astronomer's Guide.* Ontario: Camden House Publishing, 1991.

Ferris, Timothy. *Galaxies.* New York: Tabori & Chang, 1990.

Gingerich, Owen. "Messier and His Catalogue." In Mallas, John H., and Evered Kreimer, *The Messier Album*, pp. 1–16. Cambridge, Mass.: Sky Publishing Corp.; Cambridge: Cambridge University Press, 1978.

Harrington, Philip S. *Touring the Universe through Binoculars.* New York: John Wiley & Sons, 1990.

Hynes, Steven, and Brent Archinal. *Star Clusters.* Richmond: Willmann-Bell, 1996.

Jones, Kenneth Glyn. *Messier's Nebulae & Star Clusters.* 2nd ed. Cambridge: Cambridge University Press, 1991.

Levy, David H. *Skywatching.* Berkeley: The Nature Company, 1994.

Luginbuhl, Christian B., and Brian A Skiff. *Observing Handbook and Catalogue of Deep-Sky Objects.* Cambridge: Cambridge University Press, 1989.

MacRobert, Alan M. *Star Hopping for Backyard Astronomers.* Cambridge, Mass.: Sky Publishing Corp., 1993.

Mallas, John H., and Evered Kreimer. *The Messier Album.* Cambridge, Mass.: Sky Publishing Corp.; Cambridge: Cambridge University Press, 1978.

Newton, Jack, and Philip Teece. *The Guide to Amateur Astronomy.* 2nd ed. Cambridge: Cambridge University Press, 1995.

Raymo, Chet. *365 Starry Nights.* New York: Prentice Hall, 1982.

Sky & Telescope magazine. Cambridge, Mass.: Sky Publishing Corp. (monthly).

Tirion, Wil. *Sky Atlas 2000.0.* Cambridge, Mass.: Sky Publishing Corp., 1981.

Tirion, Wil, Barry Rappaport, and George Lovi. *Uranometria 2000.0.* 2 vols. Richmond: Willmann-Bell, 1987.